Quantities and Units of Measurement
A Dictionary and Handbook

J. V. Drazil

An Alexandrine Press Book

MANSELL PUBLISHING LIMITED · LONDON
OSCAR BRANDSTETTER VERLAG · WIESBADEN

Mansell Publishing Limited, 6 All Saints Street,
London N1 9RL

Oscar Brandstetter Verlag GmbH & Co. KG,
Postfach 1708, D-6200 Wiesbaden

First published 1983

Distributed in the United States and Canada by The H. W. Wilson Company, 950 University Avenue, Bronx, New York 10452

This book was commissioned, edited and designed by Alexandrine Press, Oxford.

British Library Cataloguing in Publication Data

Drazil, J. V.

Quantities and units of measurement.—2nd ed.
1. Physical measurement—Dictionaries
I. Title
530.8´03´21 QC39

ISBN 0-7201-1665-1

CIP-Kurztitelaufnahme der Deutschen Bibliothek

Drazil, J. V.

Quantities and units of measurement: a dictionary and handbook/J. V. Drazil.—
1. publ.—London: Mansell; Wiesbaden: Brandstetter, 1983.
Mansell; Wiesbaden: Brandstetter, 1983.
ISBN 3-87097-117-7 (Brandstetter)
ISBN 0-7201-1665-1
NE: HST

Text set in 10/12pt Times and printed in Great Britain by Henry Ling Limited, Dorchester, and bound by Butler and Tanner, Frome

Contents

Acknowledgements

I wish to express my thanks to Dr A. Kučera who gave me the impetus to prepare this book and assisted me during its writing, and Dr Keith Stead for his helpful reading of both the typescript and the proofs.

Much information in this book has been based on data obtained from ISO Standards, British Standards and other standards written in English, French and German (see Bibliography) by permission of the British Standards Institution, 2 Park Street, London W1A 2BS from whom copies of the complete publications may be obtained.

The values of constants are taken from the Report of the CODATA Task Group on Fundamental Constants (August 1973). I am indebted to E. Richard Cohen, chairman of the Group, for valuable information on the latest developments in the field of constants gleaned from the manuscripts of his articles, with the copies of which he kindly presented me.

The librarians of the Buckinghamshire County Libraries in Aylesbury and Amersham, of the British Standards Institution Library and of the Science Reference Library deserve my special gratitude for their assistance.

Finally I would like to thank Messrs Saunders & Dolleymore for providing facilities for the preparation of the typescript.

Introduction

Correct use and sound knowledge of units of measurement, quantities, constants and their symbols are of prime importance to all those working in science and technology, whether they be students, or professionals in academic institutions or in industry.

This book is intended to provide the fundamental information needed to meet these requirements. It is subtitled *A Dictionary and Handbook* to indicate its character: a *dictionary* because the entries are condensed and arranged in alphabetical order, and a *handbook* because it explains the relationship between the four groups of entries, namely units and their symbols and quantities and their symbols.

It is divided into three parts. Part 1 comprises an alphabetical listing of the names, symbols and abbreviations of units of the International System of units (SI units) and other commonly used units (including all-important compound units), giving their usage and conversion factors. Part 2 gives, again in alphabetical order, the quantities and constants used in all major fields of science and technology, their symbols, French and German names, dimensions, and the SI units used for their measurement. Part 3 is a list of the symbols used to denote quantities and constants.

Ideally everybody should know all commonly used units of measurement and quantities and their interrelationship, but that is near impossible. The cross-referencing employed throughout this text enables the user with only part of the information about a unit or quantity (or its symbol) to obtain all the missing information. Because all the names of quantities and constants (and where necessary also units) are given in English, French and German, someone knowing only one of the languages can, with the help of the cross-referencing and the French and German indexes, make full use of the book.

The following sections give guidance on the use of this book, explaining the entries contained in Parts 1, 2 and 3, and the symbols and abbreviations employed throughout the text. These are followed by a section on systems and units, and also the expressions used in the names of units and the prefixes and symbols used for their decimal multiples and sub-multiples in English, French and German.

Further information is included in the appendices, namely, the definitions of the base SI units; UK and US units presented in tabular form; the recommended consistent values of the fundamental constants; and a bibliography of both international and national standards relating to units and quantities.

The sources from which information in this book is drawn are noted in the Acknowledgements, the Bibliography, and elsewhere in the text as appropriate.

Throughout the book English spelling is used (which is also the spelling used in ISO and IEC publications).

The publishers and author will be grateful for any suggestions for improvement of this book.

Entries in Part 1

For practical reasons, the alphabetical arrangement used both here and in Part 2 is that commonly used in telephone directories. It has only two rules:

(*a*) a hyphenated word is deemed to be a single word;

(*b*) a space between two words is deemed to be a letter preceding 'a' in the alphabet.

Thus, for instance, 'pound inch squared' precedes 'poundal', which in turn precedes 'pound-force'.

With some exceptions, units which are of historical interest only, and also units which are not or cannot be converted into SI units (such as various units used in hydrometry, viscosimetry, for measuring hardness, photographic emulsion speed, etc.) are not included in this book. On the other hand, several units are included for which there is no corresponding quantity in Part 2; this is mostly in cases where no symbol for that quantity has been formally adopted internationally.

Multiples and sub-multiples of various single units are apparent from the examples under the individual entries of the relevant SI prefixes contained in this Part.

Entry for a Unit

The entry for a unit comprises:

1. The *name* of the unit.

2. The *symbol* used for the unit, which for an SI unit is an SI symbol and for other units is either

(*a*) an *internationally adopted symbol* for the unit (if there is such a symbol); or

(*b*) a *commonly used symbol* for the unit. The latter symbol is always in parentheses to distinguish it from the former. If (*c*) *no symbol* is commonly used for the unit, a dash in parentheses (—) follows the name of the unit.

3. *The French* and *German names* of the unit. This applies only to the more important units having their own name if that name differs significantly from the English one. The names of other units can be derived from the table 'Expressions Used in the Names of Units' on page 16, and the table 'SI Prefixes and their Symbols' on page 17.

4. The *system* or *country* in which the unit is (or was) used. If a unit is used in more than one system, only the most important system is mentioned.

5. A *comment,* such as obsolete or deprecated, where appropriate. Not all obsolete or deprecated units are so marked as a unit deprecated in one country may be used in another. It is advisable to use only SI and oSI units, and their multiples or sub-multiples formed by SI prefixes (or their symbols).

6. The *quantity* or *quantities* for which a unit is used. With a few exceptions, only one quantity is given for non-SI units.

7. Indication of the *relation* of the unit (if it is an SI unit) to other SI units

(*a*) by showing its *'dimension',* i.e. its reduction to base SI units expressed by multiplication (without any multiplication symbol – cf. dimensions of quantities) of base SI units in the order shown on p. 12 raised to positive or negative powers, where necessary; and

(*b*) by indicating other *SI unit(s) having the same 'dimension'* expressed by multiplication or division (with multiplication and division symbols) of SI units raised to positive powers, where necessary, (to negative powers only if it is not possible to avoid it). For example,

coulomb per kilogram $\mathrm{C/kg = kg^{-1}\,s\,A = A{\cdot}m^2/(J{\cdot}s) = T^{-1}{\cdot}s^{-1}}$

In simple cases only one of the two relations is shown.

8. *Conversion factor(s)* for the conversion of the unit to other unit(s) for the same quantity. In the case of non-SI units the first conversion factor is the conversion to the appropriate SI unit. All conversion factors printed in bold type are exact values.

NOTE: Scientific notation is used throughout, in which, with a very few exceptions, multiplication by $10^{\pm n}$ is used wherever possible. This is believed to be more convenient than multiplication by $10^{\pm 3n}$. Some conversion factors are accompanied by an indication of their calculation in which, to save space, no units are shown.

9. Indication of Tables (see pages 279 to 283) where *multiples and/or submultiples* (M/S) of UK and US units, not formed by SI prefixes, may be found.

10. A *note* (where appropriate).

Entry for a Symbol or Abbreviation

The entry for a symbol or an abbreviation of a unit comprises:

1. The *symbol* or *abbreviation.*

2. The *country* or *language* in which the abbreviation is used and the *full name* in the language of the country (where appropriate).

3. A *comment,* such as obsolete or deprecated, where appropriate.

4. The *name* of the unit (in English) for which the symbol or abbreviation is used.

5. The *correct symbol* (in parentheses) in cases where the entry symbol is an incorrect one.

Entries in Part 2

Great care has been taken to ensure that the names of quantities in English, French and German correspond to those recommended by international and national standards. Some alternative or obsolescent terms are included to assist readers who may meet them in literature or who are not acquainted with the latest trends in this field.

Entry for a Quantity or Constant

The entry for a quantity or constant comprises:

1. The *name* of the quantity or constant in English, French and German. Names appearing in the latest editions of ISO Standards and of British, French and German Standards (see Bibliography) are used in preference to others.

2. The *symbol* for the quantity or constant recommended by the ISO.

Naturally, names and symbols not appearing in any of the mentioned Standards are also included for various reasons. To assist the reader ISO reserve symbols and symbols not appearing in ISO publications are in parentheses. To further indicate preferred use of names and symbols a comma (,) is used to show that a name or symbol after it is on a par with that before it, and a semicolon (;) is used to show that the name or symbol before it is preferred to that after it. Also names and symbols not in parentheses are preferred to those in parentheses.

3. The *dimension* of the quantity or constant. The dimensional symbols are:

L	length	Θ	temperature
M	mass	N	amount of substance
T	time	J	luminous intensity
I	electric current		

4. *SI unit* for the quantity or constant.

5. *Other unit* which may be used in addition to, or instead of, an SI unit.

6. *Multiples and/or sub-multiples* (M/S) of the SI unit (when used for that quantity) recommended by ISO. Such recommendations exists in a few cases only.

7. (For constants) the latest *value* recommended by the CODATA Task Group on Fundamental Constants (see Acknowledgements) without the standard deviations (which can be found in Appendix 3).

8. A *note* (where appropriate).

Symbols and Abbreviations Used

1. Mathematical Symbols

.	decimal point (for example, 3.141 59)
$\times$	multiplied by (for example, 3.6×10^3)
/	divided by; per (for example, 1/16 or m/s)
=	(exactly) equal to
lb	$\log_2$
lg	$\log_{10}$
ln	$\log_e$

NOTES

(*a*) The decimal point is still used in English speaking countries and consequently in this book, although in all other countries and even in ISO Standards written in English the decimal comma (,) is used.

(*b*) Multiplication of numbers is indicated by the symbol $\times$; multiplication of symbols of quantities is indicated by the centred dot (·), for example, Pa·s; and multiplication of dimensional symbols is indicated by mere positioning of the symbols next to each other, for example, LTI.

(*c*) The symbol = is used to mean both 'exactly equal' and 'approximately equal'. This is possible because all exact numerical values are printed in bold type. Also for simplicity the symbol 'equals' (=) is used, although 'corresponds to' ($\hat{=}$) would have been more accurate.

2. General Abbreviations

abbr.	abbreviation
BRD	Federal Republic of Germany
cf.	compare

DDR	German Democratic Republic
Def	definition
depr.	deprecated
DIM	dimension
e.g.	for example
F	French
G	German
i.e.	that is
M/S	multiples and/or sub-multiples
obsol.	obsolete; obsolescent
q.v.	which see
UK	United Kingdom
US	United States

3. Relationship of Units to Systems and Countries

CGS unit	unit of the cm-g-s system
CGSB unit	unit of the cm-g-s-Bi system
CGSe unit	unit of the electrostatic CGS system; esu unit
CGSF unit	unit of the cm-g-s-Fr system
CGSm unit	unit of the electromagnetic CGS system; emu unit
FPS unit	unit of the ft-lb-s system
ft-lbf-s unit	unit of the ft-lbf-s system
m-kgf-s unit	unit of the m-kgf-s system
m-kp-s unit	unit of the m-kp-s system
MTS unit	unit of the m-t-s system
mSI unit	unit which is a multiple or sub-multiple of an SI unit
oSI unit	other unit which can be used in addition to SI units
SI unit	unit of the SI system (International System)

UK unit	unit used in the United Kingdom
US unit	unit used in the United States

NOTE

When a unit is a unit of the SI system and also of some other system, the other system is mentioned only exceptionally.

4. Other Useful Abbreviations

AEF	Normenausschuß Einheiten und Formelgrößen im DIN
AFNOR	Association Française de Normalisation
ANSI	American National Standards Institute
ASA	American Standards Association
ASMW	Amt für Standardisierung, Meßwesen und Warenprüfung (DDR)
BIH	Bureau International de l'Heure
BIPM	Bureau International des Poids et Mesures
BSI	British Standards Institution
CCU	Consultative Committee for Units (of the CIPM)
CEN	Comité Européen de Normalisation
CENELEC	Comité Européen de Normalisation Électrotechnique
CGPM	Conférence Générale des Poids et Mesures
CIE	Commission Internationale de l'Éclairage
CIPM	Comité International des Poids et Mesures
CODATA	Committee on Data for Science and Technology (of the ICSU)
CSA	Canadian Standards Association
DIN	Deutsches Institut für Normung e.V. (BRD)
DNA	Deutscher Normenausschuß e.V. (BRD)
IAU	International Astronomical Union
IBN	Institut Belge de Normalisation
ICPS	International Conference on Properties of Steam
ICRU	International Commission on Radiological Units and Measurements
ICSU	International Council of Scientific Unions

IEC	International Electrotechnical Commission
IIRS	Institute for Industrial Research and Standards (Ireland)
ISI	Indian Standards Institution
ISO	International Organization for Standardization
IUPAC	International Union of Pure and Applied Chemistry
IUPAP	International Union of Pure and Applied Physics
NBS	National Bureau of Standards (USA)
NPL	National Physical Laboratory (UK)
OIML	Organisation Internationale de Métrologie Légale
OIPM	Organisation Internationale des Poids et Mesures
ON	Österreichisches Normungsinstitut
PSI	Pakistan Standards Institution
SAA	Standards Association of Australia
SABS	South African Bureau of Standards
SANZ	Standards Association of New Zealand
SCC	Standards Council of Canada
SNV	Schweizerische Normen-Vereinigung
UMR	The Units of Measurement Regulations (UK)
USASI	USA Standards Institute
WMA	Weights and Measures Act (UK)

Systems and Units

1. International System of Units

The International System of Units, abbreviated SI in all languages, was adopted by the 11th CGPM in 1960. In its present form the System is based on seven *base units* for the seven *base quantities* listed below:

BASE QUANTITY	BASE SI UNIT	SYMBOL
length	metre	m
mass	kilogram	kg
time	second	s
electric current	ampere	A
thermodynamic temperature	kelvin	K
amount of substance	mole	mol
luminous intensity	candela	cd

The System further comprises two *supplementary SI units* (radian and steradian) and a number of *derived SI units.* In October 1980 the CIPM decided to interpret the supplementary units as dimensionless derived units, the possibility of using them or not in the expressions of derived units having been left open. Derived SI units are formed from base SI units and supplementary SI units by multiplication and division. Some derived SI units have special names, and can be used in the formation of other derived SI units.

The supplementary units, and derived units having special names and symbols and approved by the CGMP, are:

QUANTITY	SI UNIT	SYMBOL
activity	becquerel	Bq
electric charge	coulomb	C
Celsius temperature	degree Celsius	°C
capacitance	farad	F
absorbed dose	gray	Gy
inductance	henry	H
frequency	hertz	Hz
energy	joule	J
luminous flux	lumen	lm
illuminance	lux	lx
force	newton	N
resistance	ohm	Ω
pressure	pascal	Pa
plane angle	radian	rad
conductance	siemens	S
dose equivalent	sievert	Sv
solid angle	steradian	sr
magnetic flux density	tesla	T
electric potential	volt	V
power	watt	W
magnetic flux	weber	Wb

The unit degree Celsius is not, strictly speaking, a derived SI unit, but may be treated as such for the purposes of this book. Base SI units, supplementary SI units and derived SI units are known collectively as *SI units.* Derived units without a special name are *compound units.*

2. Additional Units

Other units, which are not SI units but were retained by the CIPM for their importance, are the following:

QUANTITY	UNIT	SYMBOL
time	minute	min
	hour	h
	day	d
plane angle	degree	...°
	minute	...′
	second	...″
volume	litre	l, L
mass	tonne	t
energy	electronvolt	eV
mass of an atom	atomic mass unit	u
length	astronomic unit	(AU)
	parsec	pc

These units, and compound units including these units only or these units and SI units, are indicated in this book as oSI units.

Other units are still occasionally used instead of or with SI units or oSI units and are tolerated to a varying degree. They include:

are	gon	neper	sone
bar	grade	octave	tex
barn	hectare	phon	var
curie	knot (internat.)	revolution	week
decibel	nautical mile (internat.)	röntgen	year
dioptre			

Still further units are used in some countries, for example the United Kingdom and the United States – see Imperial units and US Customary units in Part 1.

3. Other Systems and Units

Apart from SI units, the more important units of the following systems are included:

Systems of Units of Mechanics

1. Absolute or LMT systems (base quantities: length, mass, time)

MKS system (base units: metre, kilogram, second)

MTS system (base units: metre, tonne, second)

CGS system (base units: centimetre, gram, second)

FPS system (base units: foot, pound, second)

2. Gravitational or LFT systems (base quantities: length, force, time)

m-kgf-s system (base units: metre, kilogram-force, second)

m-kp-s system (base units: metre, kilopond, second)

ft-lbs-s system (base units: foot, pound-force, second)

Systems of Units of Electricity and Magnetism

MKSA system (base units: metre, kilogram, second, ampere)

CGSm system (electromagnetic CGS system)

CGSe system (electrostatic CGS system)

CGSF system (base units: centimetre, gram, second, franklin)

CGSB system (base units: centimetre, gram, second, biot)

Units of other systems and units which do not belong to any system are also included in this dictionary.

NOTES

(*a*) MKS and MKSA systems (and also MKS°K and m-sr-s-cd systems) form part of the SI system and are therefore not referred to separately.

(*b*) m-kgf-s and m-kp-s systems are identical (they only use a different name for the same unit of force), and m-kp-s system is therefore not referred to separately; m-kgf-s units can be changed to m-kp-s units by simply substituting kp for kgf.

Expressions Used in the Names of Units

ENGLISH	FRENCH	GERMAN
square ...	... carré(e)	Quadrat ...
... squared	... carré(e)	Quadrat ... (... hoch zwei)
cubic ...	... cube	Kubik ...
... cubed	... cube	Kubik ...; (... hoch drei)
... to the fourth power	... à la puissance quatre	... hoch vier
reciprocal ...	par ...; ... inverse	reziproke (-r, -s)
... per ...	... par ...	... durch; ... je ...

NOTES

(*a*) In English, within the SI, the modifiers 'square' and 'cubic' are used only with the metre. With all other units, and in a few instances also with the metre (where no area or volume is involved), the modifiers 'squared' and 'cubed' are used. The same applies to all multiples and sub-multiples of said units.

(*b*) Also the term 'Sekundenquadrat' is used in German.

(*c*) The terms '... hoch zwei' and '... hoch drei' are rarely used.

(*d*) The term 'mètre bicarré' is also used.

(*e*) The terms 'reciprocal ...' (including 'reciprocal square metre' and 'reciprocal cubic metre') are used in this book. Alternative terms are '... to the power minus one (two, three)'. In addition to the standard French terms '... à la puissance moins un (deux, trois)' also 'par ...' (e.g. 'par henry' or 'par seconde') and '... inverse' (e.g. 'henry inverse') are sometimes used. 'Eins je ...' is used in the DDR.

(*f*) The German Committee for Units (Normenausschuß Einheiten und Formelgrößen im DIN) recommends that only '... durch ...' be used in the name of units, and not '... je ...' used formerly in German standards, or '... pro ...' used in older books. Austrian and DDR standards use '... je ...'.

SI Prefixes and their Symbols for Decimal Multiples and Sub-Multiples of Units

PREFIX English	French	German	SYMBOL	FACTOR BY WHICH THE UNIT IS MULTIPLIED
exa	exa	Exa	E	10^{18}
peta	peta	Peta	P	10^{15}
tera	téra	Tera	T	10^{12}
giga	giga	Giga	G	10^{9}
mega	méga	Mega	M	10^{6}
kilo	kilo	Kilo	k	10^{3}
hecto	hecto	Hekto	h	10^{2}
deca	déca	Deka	da	10
deci	déci	Dezi	d	10^{-1}
centi	centi	Zenti	c	10^{-2}
milli	milli	Milli	m	10^{-3}
micro	micro	Mikro	μ	10^{-6}
nano	nano	Nano	n	10^{-9}
pico	pico	Piko	p	10^{-12}
femto	femto	Femto	f	10^{-15}
atto	atto	Atto	a	10^{-18}

NOTES

(*a*) SI prefixes and their symbols can be used, in theory, with all SI units having special names, but in practice are not used with the steradian and degree Celsius and rarely with the candela and the kelvin. They may also be used with most oSI units and some non-SI units but, with one or two exceptions, not with units having non-decimal multiples and sub-multiples (e.g. units of time and plane angle, Imperial units and US Customary units) and some other units (e.g. knot and nautical mile).

(*b*) The prefixes (symbols) are directly attached to the names (symbols) of units.

(*c*) Multiples and sub-multiples of units formed by prefixes (symbols) can be raised to positive or negative powers, e.g.

$$(1\ \mathrm{mm})^2 = (10^{-3}\ \mathrm{m})^2 = 10^{-6}\ \mathrm{m}^2$$

(*d*) Prefixes hecto, deca, deci and centi should be avoided as far as possible.

(*e*) No other prefixes (e.g. myria) or compound prefixes (e.g. decimilli) may be used.

(*f*) The gram and the symbol g (rather than the kilogram and kg) are used as a basis for the formation of names and symbols of the multiples and sub-multiples of the unit of mass.

Part 1

A Dictionary of Units of Measurement

Their Symbols and Abbreviations

A

a are, year; atto; ab

A ampere; atta

Å ångström

aA abampere

ab (a)
a prefix denoting a CGSm unit (used in US)

abampere (aA)
CGSm unit of electric current
10 A

abampere centimetre squared ($aA{\cdot}cm^2$)
CGSm unit of electromagnetic moment
10^{-3} $A{\cdot}m^2$

abampere per square centimetre (aA/cm^2)
CGSm unit of current density
10^5 A/m^2

abcoulomb (aC)
CGSm unit of electric charge
10 C

abcoulomb centimetre ($aC{\cdot}cm$)
CGSm unit of electric dipole moment
10^{-1} $C{\cdot}m$

abcoulomb per cubic centimetre (aC/cm^3)
CGSm unit of volume density of charge
10^7 C/m^3

abcoulomb per square centimetre (aC/cm^2)
CGSm unit of electric polarization and electric flux density
For electric polarization:
10^5 C/m^2
For electric flux density:
$7.957\,75 \times 10^3$ C/m^2 $[=10^5/(4\pi)]$

abfarad (aF)
CGSm unit of capacitance
10^9 F

abhenry (aH)
CGSm unit of inductance
10^{-9} H

abmho (a℧)
=absiemens (q.v.)

abohm (aΩ)
CGSm unit of resistance
10^{-9} Ω

abohm centimetre (aΩ·cm)
CGSm unit of resistivity
10^{-11} Ω·m

abs see: absolute

absiemens (aS)
CGSm unit of conductance
10^9 S

absiemens per centimetre (aS/cm)
CGSm unit of conductivity
10^{11} S/m

absolute
Note: depr. adjective sometimes used to distinguish MKSA(=SI) units from the so-called international units (obsol. since 1947), so that e.g. the symbol A_{abs} is used for the ampere (A) to distinguish it from the international ampere (A_{int})

abstat depr. prefix denoting a CGSe unit

abtesla (aT)
=gauss (q.v.)

abvolt (aV)
CGSm unit of electric potential
10^{-8} V

abvolt per centimetre (aV/cm)
CGSm unit of electric field strength
10^{-6} V/m

abweber (aWb)
= maxwell (q.v.)

aC abcoulomb

acoustic ohm (—)
depr. unit of acoustic impedance
10^5 Pa·s/m^3
Note: name sometimes given to dyne second per centimetre to the fifth power

acre (–)
UK and US units of area
$4.046\,86 \times 10^3$ m^2 (= **4046.856 422 4** m^2)
$4.046\,86 \times 10^{-1}$ ha
M/S: Table 2
Note: used for agrarian measurements

acre per pound (acre/lb)
UK and US unit of specific surface
$8.921\,79 \times 10^{-1}$ ha/kg

acre-foot (acre·ft)
US unit of volume
$1.233\,48 \times 10^3$ m^3
4.356 $\times$ **10^4** ft^3
Note: used in irrigation engineering

acre-foot per day (acre·ft/d)
US unit of volume flow rate
$1.427\,64 \times 10^{-2}$ m^3/s
$1.233\,48 \times 10^3$ m^3/d
4.356 $\times$ **10^4** ft^3/d

acre-foot per hour (acre·ft/h)
US unit of volume flow rate
$1.233\,48 \times 10^3$ m³/h
4.356 $\times$ **10⁴** ft³/h

acre-inch (acre·in)
US unit of volume
$1.027\,90 \times 10^2$ m³
3.630 $\times$ **10³** ft³

admiralty mile obsol. name for nautical mile (UK)

AE G symb. for *astronomische Einheit* = astronomical unit

aF abfarad

A.G. G abbr. for *Atomgewicht* = atomic weight

ah: a-h ampere hour (A·h)

aH abhenry

Ah ampere hour (A·h)

a.l. F abbr. for *année de lumière* = light year

amp, Amp ampere (A)

ampere A
ampère
Ampere
base SI unit of electric current and SI unit of current linkage, magnetic potential difference and magnetomotive force
Def: see Appendix 1

ampere × circular mil (A × circular mil)
UK and US unit of electromagnetic moment
$5.067\,07 \times 10^{-10}$ A·m²

ampere hour A·h
oSI unit of electric charge
3.6×10^{3} C or A·s

ampere metre squared A·m^{2}
SI unit of electromagnetic moment
Note: cf. ampere square metre

ampere minute A·min
oSI unit of electric charge
60 A·s

ampere per inch A/in
UK and US unit of magnetic field strength
3.397 01×10 A/m

ampere per kilogram A/kg
=coulomb per kilogram second (q.v.)

ampere per metre A/m
SI unit of magnetic field strength, (lower, upper and thermodynamic) field strength, magnetization and linear current density
=N/Wb

ampere per square inch A/in^{2}
unit of current density
1.550 00×10^{3} A/m^{2}

ampere per square metre A/m^{2}
SI unit of current density

ampere per square metre kelvin squared A/(m^{2}·K^{2})
SI unit of Richardson constant

ampere per volt A/V
=siemens (q.v.)

ampere per weber A/Wb
=reciprocal henry (q.v.)

ampere second A·s
=coulomb (q.v.)

ampere square metre $A \cdot m^2$
SI unit of magnetic moment of a particle, Bohr magneton and nuclear magneton
$= m^2 A^{-1} = J/T$
Note: cf. ampere metre squared

ampere square metre per joule second $A \cdot m^2/(J \cdot s)$
SI unit of gyromagnetic coefficient
$= C/kg = s^{-1} \cdot T^{-1}$

ampere-turn (At)
ampère-tour (*At*)
Amperewindung (*Aw*)
obsol. unit of magnetomotive force
Def: number of turns of a coil times current (in amperes) flowing through the coil

ampere-turn per metre (At/m)
=ampere per metre (q.v.)

amu; a.m.u. atomic mass unit (old physical)

ångström Å
ångström
Ångström
unit of wavelength
$\mathbf{10}^{-10}$ m = $\mathbf{10}^{-1}$ nm

ap; ap. apothecaries'

api ampere per inch (A/in)

apostilb asb
unit of luminance
$3.183\ 10 \times 10^{-1}$ cd/m² (=1/π)
Note: formerly also called blondel

apoth; apoth. apothecaries'

apothecaries' drachm, dram, ounce, pound
see: drachm, dram, ounce, pound (apothecaries')

apothecaries' units
obsolete units of mass formerly used in UK and US and including the following:

1 apothecaries' ounce = **24** scruples
(oz apoth in UK, oz ap in US)
1 drachm (—) (in UK)
1 dram (dr) (in US) } = **3** scruples
1 scruple (—) = **20** grains
1 grain (gr) (no symb. in US) = **1/480** apothecaries' ounce

Note: see Table 10; except for grain not lawful for trade in the UK

a.p.s.i. amperes per square inch (A/in^2)

are a
are
Ar
unit of area
$\mathbf{10}^2\ m^2 = \mathbf{10}^{-2}\ ha = \mathbf{10}^{-4}\ km^2$
$1.195\,99 \times 10^2\ yd^2$
Note: used for agrarian measurements only; the hectare (q.v.) is more widely used in practice than the are

aS absiemens

asb apostilb

assay ton (UK) (—)
unit of mass
$3.266\,67 \times 10^{-2}\ kg = 32.667\ g$ $[=(7/480) \times 2.24]$

assay ton (US) (—)
unit of mass
$2.916\,67 \times 10^{-2}\ kg = 29.167\ g$ $[=(7/480) \times 2]$

astronomic(al) unit (AU)
unité astronomique (*UA*)
astronomische Einheit (*AE*)
unit of length
$1.495\,978\,70 \times 10^{11}$ m (adopted 1979)

at technical atmosphere

At; a.t.; A.T. ampere-turn; assay ton (—)

ata G obsol. and depr. form of symbol 'at' when used for measuring absolute pressure, cf. psia

At.-Gew. G abbr. for *Atomgewicht* (=atomic weight)

atm standard atmosphere

atmosphere see: standard atmosphere; technical atmosphere

at. no. atomic number

atomic mass unit (old chemical) (—)
obsol. unit of atomic mass ('weight')

Def: 1 atomic mass unit = 1/16 of the mass of an atom of oxygen of the naturally occurring isotopic composition

= 1 u/1.000 043

Note: also other values were used; also called atomic weight unit

atomic mass unit (old physical) (amu)
obsol. unit of atomic mass ('weight')

Def: 1 amu = 1/16 of the mass of an atom of nuclide ^{16}O
= 1 u/1.000 318
Note: also other values were used

atomic mass unit (unified) (u)
unité de masse atomique (*unifiée*)
atomare Masseneinheit (*vereinheitlichte*)
oSI unit of (unified) atomic mass constant

Def: 1 u = 1/12 of the rest mass of a neutral atom of the nuclide ^{12}C in the ground state

$1.660\,565\,5 \times 10^{-27}$ kg

Note: also unit of (rest) mass of particles, mass excess and mass defect; the rider '(unified)' may be left out

atomic weight unit (awu)
Note: name sometimes used for atomic mass unit (old chemical) (q.v.)

atta (A)
depr. prefix denoting $\times 10^{18}$ – see: exa

atto a
SI prefix denoting $\times 10^{-18}$. Examples: attocoulomb (aC), attojoule (aJ), attometre (am)

atu G obsol. and depr. form of symbol 'at' when used for measuring pressure below atmospheric pressure

atü G obsol. and depr. form of symbol 'at' when used for measuring gauge pressure, cf. psig

A-turn ampere-turn

at. wt. atomic weight

AU: A.U. astronomic unit

ÅU ångström unit = ångström

aV abvolt

av avoirdupois

avdp avoirdupois

avoirdupois units

units of mass used in UK and US. They include the following:

1 ton (—)	=	**2240** lb
1 hundredweight (cwt)	=	**112** lb
1 cental (ctl)	=	**100** lb
1 quarter (qr)	=	**28** lb
1 stone (—)	=	**14** lb
1 pound (lb)	=	**0.453 592 37** kg (fundamental)
1 ounce (oz)	=	**1/16** lb
1 dram (dr)	=	**1/16** oz
1 grain (gr)	=	**1/7000** lb

Note: In the US the first two units are called long ton or gross ton, and long hundredweight, respectively, and are rarely used. Instead the following units are used:

1 short ton (—)	=	**2000** lb
1 short hundredweight (—)	=	**100** lb

Cental and stone are not used and grain has no officially approved symbol in the US. Cf. Tables 7 and 8.

A.W.; Aw G. abbr. for *Amperewindung*=ampere turn

awu atomic weight unit

aΩ abohm

B

b barn; bar (bar)

B bel; brewster

ba barye

bar bar
bar
Bar
unit of pressure (of fluid)
$\mathbf{10^5}$ Pa = $\mathbf{10^2}$ kPa = $\mathbf{10^{-1}}$ MPa
1.019 72 kgf/cm^2 or kp/cm^2
1.450 38 × 10 lbf/in^2

barn b
barn
Barn
unit of cross section
= $\mathbf{10^{-28}}$ m^2 = $\mathbf{10^2}$ fm^2 (square femtometre)

barn per electronvolt b/eV
unit of spectral cross section
6.241 46 × 10^{-10} m^2/J

barn per erg b/erg
unit of spectral cross section
$\mathbf{10^{-21}}$ m^2/J

barn per steradian b/sr
unit of angular cross section
$\mathbf{10^{-28}}$ m^2/sr

barn per steradian electronvolt b/(sr·eV)
unit of spectral angular cross section
6.241 46 × 10^{-10} m^2/(sr·J)

barn per steradian erg b/(sr·erg)
unit of spectral angular cross section
$\mathbf{10^{-21}}$ m^2/(sr·J)

barrel (—)
US unit of volume (capacity) for petroleum, etc.
1.589 87 × 10^{-1} m^3
1.589 87 × 10^2 dm^3 or litre
4.2 **× 10** gallon (US)

barrel see: dry barrel; other barrels are not included

barye (ba)
=dyne per square centimetre (q.v.)

baud (—)
unit of transmission speed or modulation rate
=reciprocal second (q.v.)

bbl dry barrel

Bd baud (—)

becquerel Bq
SI unit of activity (of radionuclide)
$1\ s^{-1} = 2.702\ 70 \times 10^{-11}$ Ci [$= 1/(3.7 \times 10^{10})$]

becquerel per cubic metre Bq/m^3
SI unit of volume activity (of radionuclide)
$= m^{-3}\ s^{-1}$

becquerel per kilogram Bq/kg
SI unit of specific activity (of a radionuclide)
$2.702\ 70 \times 10^{-11}$ Ci/kg
$= kg^{-1}\ s^{-1}$

becquerel per metre Bq/m
SI unit of linear activity (of radionuclide)
$= m^{-1}\ s^{-1}$

becquerel per mole Bq/mol
SI unit of molar activity (of radionuclide)
$= s^{-1}\ mol^{-1}$

bel
bel
Bel
Note: its sub-multiple decibel (q.v.) is used in practice

BeV depr. US symbol for billion electronvolts
$= \mathbf{10^9}$ eV $= \mathbf{1}$ GeV

bhp, B.H.P. brake horsepower; British horsepower

Bi biot

billion
$=\mathbf{10^{12}}$ in European countries, including UK (cf. tera); but the US value is being increasingly accepted in some fields (e.g. economy and finance)
$=\mathbf{10^{9}}$ in US and Canada (cf. giga)

billionth
$=\mathbf{10^{-12}}$ in European countries, including UK (cf. pico)
$=\mathbf{10^{-9}}$ in US and Canada (cf. nano)

biot (Bi)
CGSB unit of electric current
10 A

biot centimetre squared (Bi·cm²)
CGSB unit of electromagnetic moment
10^{-3} A·m²

biot per centimetre (Bi/cm)
CGSB unit of magnetic field strength
$7.957\,75 \times 10$ A/m

biot second (Bi·s)
CGSB unit of electric charge
10 C

bit (—)
binary unit of:
(a) information content (decision between two equally possible states)
(b) capacity in an information storage device (logarithm to base two of the number of possible states in the device)

bit per second bit/s
unit of bit rate or data signalling rate
=reciprocal second (q.v.)

bit per unit area (e.g. bit/mm², bit/cm², bit/in²)
unit of (surface) bit density

bit per unit length (e.g. bit/mm, bit/cm, bit/in)
unit of (linear) bit density
Note: the above two units are used as units of information density

Blindwatt (bW, BW)
G. depr. unit of reactive power (= var)

blondel (—)
= apostilb (q.v.)

board foot (—)
unit of volume (for timber only)
$2.359\,74 \times 10^{-3}$ m^3
$\mathbf{1.44} \times \mathbf{10^2}$ in^3

bougie nouvelle name of a unit of luminous intensity changed at 9th CGPM, 1948, to candela (q.v.)

bpd barrel per day

bps bits per second (bit/s)

Bq becquerel

brake horse-power (bhp)
= horsepower (q.v.)

brewster (B)
unit of stress optical coefficient
$\mathbf{10^{-12}}$ Pa^{-1} = **1** pPa^{-1}

British thermal unit Btu
obsol. unit of heat
$1.055\,06 \times 10^3$ J (= $2.326 \times 453.592\,37$)
$2.519\,96 \times 10^{-1}$ kcal$_{IT}$
$2.930\,71 \times 10^{-4}$ kW·h
$7.781\,69 \times 10^2$ ft·lbf
Note: this is the 'international table British thermal unit' adopted by the 5th ICPS, 1956. This unit is used throughout this dictionary

British thermal unit — other values (all obsol.)

Btu, mean	1055.87 J
Btu, thermochemical	1054.35 J
Btu, water	1054.54 J
Btu, 39 °F	1059.67 J
Btu, 60 °F	1054.68 J

British thermal unit foot per square foot hour degree Fahrenheit or **Rankine** Btu·ft/(ft²·h· °F) or Btu·ft/(ft²·h· °R)
unit of thermal conductivity
1.730 73 W/(m·K)

British thermal unit inch per square foot hour degree Fahrenheit or **Rankine** Btu·in/(ft²·h· °F) or Btu·in/(ft²·h· °R)
unit of thermal conductivity
$1.442\,28 \times 10^{-1}$ W/(m·K)
$1.240\,14 \times 10^{-1}$ $kcal_{IT}$/(m·h·K)

British thermal unit per cubic foot Btu/ft³
unit of calorific value (volume basis)
$3.725\,89 \times 10^{4}$ J/m³
8.899 15 $kcal_{IT}/m^3$

British thermal unit per cubic foot hour Btu/(ft³·h)
unit of heat release rate
$1.034\,97 \times 10$ W/m³
8.899 15 $kcal_{IT}/(m^3 \cdot h)$

British thermal unit per foot hour degree Fahrenheit or **Rankine** Btu/(ft·h· °F) or Btu/(ft·h· °R)
unit of thermal conductivity
1.730 73 W/(m·K)
1.488 16 $kcal_{IT}$/(m·h·K)

British thermal unit per foot second degree Fahrenheit or **Rankine** Btu/(ft·s· °F) or Btu/(ft·s· °R)
unit of thermal conductivity
$6.230\,64 \times 10^{3}$ W/(m·K)

British thermal unit per hour Btu/h
unit of heat flow rate
$2.930\,71 \times 10^{-1}$ W
$2.519\,96 \times 10^{-1}$ $kcal_{IT}$/h

British thermal unit per pound Btu/lb
unit of specific internal energy and calorific value (mass basis)
2.326 $\times$ **10³** J/kg
5.555 56 $\times 10^{-1}$ $kcal_{IT}$/kg

British thermal unit per pound degree Fahrenheit or **Rankine** Btu/(lb· °F) or Btu/(lb· °R)
unit of specific heat capacity
4.186 8 $\times$ **10³** J/(kg·K)
1 $kcal_{IT}$/(kg·K)

British thermal unit per square foot hour Btu/(ft²·h)
unit of density of heat flow rate
3.154 59 W/m²
2.712 46 $kcal_{IT}/(m^2\cdot h)$

British thermal unit per square foot hour degree Fahrenheit or **Rankine** Btu/(ft²·h· °F) or Btu/(ft²·h· °R)
unit of coefficient of heat transfer
5.678 26 W/(m²·K)
4.882 43 $kcal_{IT}$/(m·h·K)

British thermal unit per square foot second degree Fahrenheit or **Rankine** Btu/(ft²·s·F) or Btu/(ft²·s· °R)
unit of coefficient of heat transfer
2.044 17 $\times 10^4$ W/(m²·K)

B.Th.U. British thermal unit (Btu)

Btu British thermal unit

bu bushel (US)

bushel (UK) (—)
obsol. UK unit of volume (capacity)
3.636 87 $\times 10^{-2}$ m³
3.636 87 $\times$ 10 dm³ or litre
1.284 35 ft³
1.032 06 bushel (US)
M/S: Table 4
Note: there are also other bushels

bushel (US) (bu)
US unit of volume (capacity) for dry measure
$3.523\,91 \times 10^{-2}$ m³
$3.523\,91 \times 10$ dm³ or litre
$2.150\,42 \times 10^{3}$ in³ (=definition)
$9.689\,39 \times 10^{-1}$ bushel (UK)
M/S: Table 6

byte (—)
unit of capacity in an information storage device
usually = 8 bits (q.v.), in which case it is sometimes called octet

bW; BW G abbr. for *Blindwatt* (q.v.)

C

c centigrade; centi; obsol. symbol for cubic; used for ccm (cm^3), cdm (dm^3) and cmm (mm^3); obsol. symbol for cycle or curie or metric carat

C coulomb; obsol. symbol for curie

°C degree Celsius

cal calorie

cal_{15} calorie, 15 °C

cal_{20} calorie, 20 °C

cal_{IT} calorie (I.T.)

cal_{th} calorie, thermochemical

Cal obsol. abbr. for kilocalorie (kcal)

calorie (I.T.) cal_{IT} or cal

obsol. unit of heat

4.186 8 **J**

1.163 $\times$ **10**$^{-6}$ kW·h

3.968 32 $\times$ 10^{-3} Btu

Note: cf. international table calorie and kilocalorie; when the name calorie or symbol cal are used unspecified after July 1956, they refer to the international table calorie

calorie – other values (all obsol.)

calorie, defined = calorie, thermochemical (q.v.)

calorie (dietetic) = 10^3 cal_{15} = 1 $kcal_{15}$ = 4.1855 kJ

calorie, large = kilocalorie (q.v.)

calorie, mean (—) 4.190 0 J

calorie, small = calorie (in contrast to kilocalorie) – cf. also gram-calorie

calorie, thermochemical cal_{th} **4.184** J

calorie, water = calorie, 15 °C (q.v.)

calorie, 15 °C (CIPM, 1950) cal_{15} 4.185 5 J

calorie, 15 °C (NBS, 1939) cal_{15} 4.185 8 J

calorie, 20 °C cal_{20} 4.181 9 J

calorie (I.T.) per centimetre second kelvin or **degree Celsius**
cal_{IT}/(cm·s·K) or cal_{IT}/(cm·s·°C)

unit of thermal conductivity

4.186 8 $\times$ **10**^{2} W/(m·K)

calorie (I.T.) per gram cal_{IT}/g

unit of specific internal energy

4.186 8 $\times$ **10**^{3} J/kg

calorie (I.T.) per gram kelvin or **degree Celsius**
cal_{IT}/(g·K) or cal_{IT}/(g·°C)

unit of specific heat capacity and specific entropy

4.186 8 $\times$ **10**^{3} J/(kg·K)

calorie (I.T.) per kelvin or **degree Celsius**
cal_{IT}/K or cal_{IT}/°C

unit of heat capacity

4.186 8 J/K

calorie (I.T.) per second cal_{IT}/s
unit of heat flow rate
4.186 8 W

calorie (I.T.) per second centimetre kelvin or **degree Celsius** $cal_{IT}/(s \cdot cm \cdot K)$ or $cal_{IT}/(s \cdot cm \cdot {}^\circ C)$
unit of thermal conductivity
4.186 8 × 10² W/(m·K)

calorie (I.T.) per second square centimetre kelvin or **degree Celsius** $cal_{IT}/(s \cdot cm^2 \cdot K)$ or $cal_{IT}/(s \cdot cm^2 \cdot {}^\circ C)$
unit of coefficient of heat transfer
4.186 8 × 10⁴ W/(m²·K)

calorie (I.T.) per square centimetre second $cal_{IT}/(cm^2 \cdot s)$
unit of density of heat flow rate
4.186 8 × 10⁴ W/m²

calorie (I.T.) per square centimetre second kelvin or **degree Celsius** $cal_{IT}/(cm^2 \cdot s \cdot K)$ or $cal_{IT}/(cm^2 \cdot s \cdot {}^\circ C)$
unit of coefficient of heat transfer
4.186 8 × 10⁴ W/(m²·K)

candela cd
candela
Candela
base SI unit of luminous intensity
Def: see Appendix 1

candela per square centimetre cd/cm²
mSI unit of luminance
10⁴ cd/m²
Note: formerly called stilb

candela per square foot cd/ft²
unit of luminance
1.076 39 × 10 cd/m²

candela per square inch cd/in²
unit of luminance
1.550 00 × 10³ cd/m²

candela per square metre cd/m^2
SI unit of luminance
1.0 × **10**$^{-4}$ cd/cm^2 or sb (stilb)
3.141 59 asb or blondel
3.141 59 × 10^{-4} La (lambert)
Note: formerly called nit

candle; new candle
obsol. name for candela (q.v.)
Note: see also bougie nouvelle, Hefner candle, and international candle
candle-foot – see: foot-candle

carat (c)
measure of purity of gold
$\frac{1}{24}$ of gold
Note: usually spelled karat in the US

carat see: metric carat

carcel (—)
obsol. unit of luminous intensity
0.98 cd

cbm G abbr. for *Kubikmeter* = cubic metre (m^3)

...^{cc} one hundredth of a centigrade (centesimal second)

cc; ccm cubic centimetre (cm^3)

cd candela

cent (—)
(a) unit of dimensionless quantity: frequency interval
The interval between two frequencies f_1, f_2 having a ratio $f_1/f_2 = \sqrt[1200]{2} = 1.000\ 58$
$= 1200$ lb (f_1/f_2)
$= 8.333\ 33 \times 10^{-4}$ octave [=(1/1200)]
(b) unit of dimensionless quantity: reactivity
= **10**$^{-2}$ dollar (q.v.)

cental (ctl)
UK unit of mass
$4.535\,92 \times 10$ kg
10^{2} lb

centesimal minute $\ldots .^{cg}$
unit of plane angle
10^{-2} grade $= 1.570\,80 \times 10^{-4}$ rad
Note: name sometimes used for centigrade (q.v.)

centesimal second $\ldots^{cc}$
unit of plane angle
10^{-4} grade $= 1.570\,809 \times 10^{-6}$ rad
Note: name sometimes used for one hundredth of a centigrade

centi c
SI prefix denoting $\times 10^{-2}$. It should be avoided as far as possible and used only where well established in practice. Examples: centigrade ($\ldots^{cg}$), centigram (cg), centilitre (cl, cL), centimetre (cm), centipoise (cP), centistokes (cSt)

centiare ca
unit of area
1 m^{2}
Note: should be avoided

centigrade $\ldots .^{cg}$
centigrade
Zentigon; (*Neuminute*)
unit of plane angle
10^{-2} grade $= 1.570\,80 \times 10^{-4}$ rad
Note: also called centesimal minute (q.v.)

centigrade; one hundredth of a ~ $\ldots .^{cc}$
unit of plane angle
10^{-4} grade $= 1.570\,80 \times 10^{-6}$ rad
Note: also called centesimal second (q.v.)

centigrade, degree
see: degree centigrade

Centigrade heat unit (CHU)
unit of heat
1.8 British thermal unit (q.v.)

centimetre cm
centimètre
Zentimeter
mSI unit and CGS base unit of length
10^{-2} m
Note: this unit is used as a fundamental unit in many branches of industry, such as building and textile industry; for other quantities see: metre; also used as a CGSe unit of capacitance (=statfarad) and a CGSm unit of inductance (=abhenry)

centimetre per second squared cm/s^2
CGS unit of (linear) acceleration
10^{-2} m/s^2
Note: also called galileo or gal

centimetre second degree Celsius per calorie (I.T.)
$cm \cdot s \cdot °C/cal_{IT}$
unit of thermal resistivity
$2.388\,46 \times 10^{-3}$ m·K/W
$2.777\,78 \times 10^{-4}$ $m \cdot h \cdot °C/kcal_{IT}$
$4.133\,79 \times 10^{-3}$ ft·h·°F/Btu

centimetre squared per second cm^2/s
=stokes (q.v.)

centimetre to the fourth power cm^4
CGS unit of second moment of area
10^{-8} metre to the fourth power (q.v.)

centipoise cP
unit of (dynamic) viscosity
10^{-3} Pa·s

centistokes cSt
unit of kinematic viscosity
10^{-6} metre squared per second (q.v.)

cfm cubic foot per minute (ft^3/min)

cfs cubic foot per second (ft^3/s)

....^{cg} centigon

cg centigram

CGSe unit (or **CGS-esu**)
unit of the so-called electrostatic CGS system. See: under stat- (e.g. statampere) or under CGSe unit (e.g. CGSe unit of permittivity)

CGSe unit of magnetic field strength
$2.654\,42 \times 10^{-9}$ A/m

CGSe unit of magnetic flux
$2.997\,92 \times 10^{2}$ Wb

CGSe unit of magnetic flux density
$2.997\,92 \times 10^{6}$ T

CGSe unit of magnetic polarization
$3.767\,30 \times 10^{7}$ T

CGSe unit of magnetization
$2.997\,92 \times 10^{7}$ A/m

CGSe unit of reluctance
$8.854\,19 \times 10^{-14}$ H^{-1}

CGSm unit (or **CGS-emu**)
unit of the so-called electromagnetic CGS system. See: under ab- (e.g. abampere) or under CGSm unit (e.g. CGSm unit of permittivity)

CGSm unit of magnetic field strength
= oersted (q.v.)

CGSm unit of magnetic flux
= maxwell (q.v.)

CGSm unit of magnetic polarization
$1.256\,64 \times 10^{-3}$ T

CGSm unit of magnetization
$\mathbf{10^{-3}}$ A/m

CGSm unit of reluctance
$7.957\,75 \times 10^{7}$ H^{-1}

ch chain; F abbr. for *cheval vapeur* = metric horsepower

chain (—)
UK and US unit of length
$\mathbf{2.011\,68 \times 10}$ m
M/S: Table 1
Note: also called Gunter's chain, imperial chain or surveyor's chain. Symbol 'ch' is used in US; see also engineer's chain

charrière (—)
F unit for grading the sizes (*diametres*) of catheters and probes
$\frac{1}{3}$ mm
Note: used within the range of $\frac{1}{3}$ mm to 10 mm; determined on *Charrière filière,* hence the name.
The symbol Fr (for French scale) is sometimes used for this unit

cheval vapeur see horsepower (metric)

CHU Centigrade heat unit

Ci curie

circular inch (—)
UK and US unit of area
$5.067\,07 \times 10^{-4}$ m^2
$7.853\,98 \times 10^{-1}$ in^2 (=$\pi/4$)
1.0 $\times \mathbf{10^{6}}$ circular mils
Note: 1 circular inch = the area of a circle 1 inch in diameter

circular mil (—)
UK and US unit of area
$5.067\,07 \times 10^{-10}$ m²
$7.853\,98 \times 10^{-7}$ in²
1.0 $\times$ **10**$^{-6}$ circular inch
Note: 1 circular mil = the area of a circle one-thousandth of an inch in diameter; cf. mil

cl, cL centilitre

clausius (—)
obsol. unit of entropy
4.186 8 $\times$ **10**^{3} J/K = **1** $kcal_{IT}/K$

clusec (—)
unit of leak rate used in vacuum technology
$1.333\,22 \times 10^{-6}$ N·m/s or W
Note: see lusec

cm centimetre

CM metric carat (—)

cont hp Continental horse-power
= metric horsepower (—)

cord (—)
unit of volume (for timber only)
3.624 56 m³
1.28 $\times$ **10**^{2} ft³

coulomb C
coulomb
Coulomb
SI unit of electric charge, electric flux and elementary charge
= s A
$2.777\,78 \times 10^{-4}$ A·h

coulomb metre C·m
SI unit of electric dipole moment
= m s A

coulomb metre squared per kilogram C·m²/kg
SI unit of specific gamma ray constant
$= m^2\ kg^{-1}\ s\ A$

coulomb metre squared per volt C·m²/V
SI unit of polarizability of molecule
$= kg^{-1}\ s^4\ A^2 = F \cdot m^2$

coulomb per cubic metre C/m^3
SI unit of volume density of charge
$= m^{-3}\ s\ A$

coulomb per kilogram C/kg
SI unit of exposure
$= kg^{-1}\ s\ A = A \cdot m^2/(J \cdot s) = T^{-1} \cdot s^{-1}$
$3.875\,97 \times 10^3$ R (röntgen)

coulomb per kilogram second C/(kg·s)
SI unit of exposure rate
$= kg^{-1} A$
$3.875\,97 \times 10^3$ R/s (röntgen per second)

coulomb per mole C/mol
SI unit of Faraday constant

coulomb per square metre C/m^2
SI unit of surface density of charge, electric flux density and electric polarization
$= m^{-2}\ s\ A$

cP centipose

cps cycle(s) per second = hertz (Hz); characters per second

crocodile (—)
depr. unit of electric potential
10^6 V = **1** MV

cSt centistokes

c/s cycle(s) per second = hertz (Hz)

ctl cental

cu; cu. depr. UK and US abbr. for cubic (e.g. cu ft = ft^3)

cubic centimetre cm^3
centimètre cube
Kubikzentimeter
mSI and CGS unit of volume
$\mathbf{10^{-6}}\ m^3 = \mathbf{1}$ ml or mL

cubic centimetre per gram cm^3/g
mSI and CGS unit of specific volume
$\mathbf{10^{-3}}\ m^3/kg = \mathbf{1}\ dm^3/kg$ or l/kg, L/kg or m^3/Mg or m^3/t

cubic centimetre per kilogram cm^3/kg
mSI unit of specific volume
$\mathbf{10^{-6}}\ m^3/kg$

cubic decimetre dm^3
decimètre cube
Kubikdezimeter
mSI unit of volume
$\mathbf{1.0 \times 10^{-3}}\ m^3$
1.0 litre

cubic decimetre per kilogram dm^3/kg
mSI unit of specific volume
$\mathbf{10^{-3}}\ m^3/kg$
1 m^3/t or cm^3/g or l/kg, L/kg or ml/g, mL/g

cubic foot ft^3
UK and US unit of volume
$2.831\,68 \times 10^{-2}\ m^3$
$2.831\,68 \times 10$ dm^3 or litre
M/S: Table 3

cubic foot per pound ft^3/lb
FPS unit of specific volume
$6.242\,80 \times 10^{-2}\ m^3/kg$
$\mathbf{2.240 \times 10^3}$ ft^3/UKton
$\mathbf{1.728 \times 10^3}$ in^3/lb

cubic foot per second ft^3/s
FPS unit of volume flow rate
$2.831\ 68 \times 10^{-2}\ m^3/s$

cubic foot per ton (UK) $ft^3/UKton$
UK unit of specific volume
$2.786\ 96 \times 10^{-5}\ m^3/kg$

cubic inch in^3
UK and US unit of volume
$1.638\ 706 \times 10^{-5}\ m^3$
$1.638\ 706 \times 10\ cm^3$
M/S: Table 3

cubic inch per pound in^3/lb
unit of specific volume
$3.612\ 73 \times 10^{-5}\ m^3/kg$
$5.787\ 04 \times 10^{-4}\ ft^3/lb$

cubic metre m^3
mètre cube
Kubikmeter
SI unit of volume
1.0 $\times$ **10^3** litre
1.307 95 yd^3
$3.531\ 47 \times 10$ ft^3
$6.102\ 37 \times 10^4$ in^3
$2.199\ 69 \times 10^2$ gallon (UK)
$2.641\ 72 \times 10^2$ gallon (US)
Note: cf. metre cubed

cubic metre per coulomb m^3/C
SI unit of Hall coefficient

cubic metre per hour m^3/h
oSI unit of volume flow rate
$2.777\ 78 \times 10^{-4}\ m^3/s$
$9.809\ 63 \times 10^3$ ft^3/s

cubic metre per kilogram m^3/kg

SI unit of specific volume

1.0 $\times$ **10**^{3} dm^3/kg or l/kg, L/kg or cm^3/g or ml/g, mL/g or m^3/t

1.601 85 $\times$ 10 ft^3/kg

2.767 99 $\times$ 10^4 in^3/lb

cubic metre per mole m^3/mol

SI unit of molar volume

cubic metre per second m^3/s

SI unit of volume flow rate

3.6 $\times$ **10**^{3} m^3/h

3.531 47 $\times$ 10 ft^3/s

Note: also unit of conductance (of a duct), intrinsic conductance, molecule conductance and recombination coefficient

cubic yard yd^3

UK and US unit of volume

7.645 55 $\times$ 10^{-1} m^3

M/S: Table 3

curie Ci

curie

Curie

unit of activity (of radionuclide)

3.7 $\times$ **10**^{10} Bq = **37** GBq

curie MeV Ci·MeV

obsol. unit of power (nuclear)

5.93 $\times$ 10^{-3} W

curie per cubic metre Ci/m^3

unit of volume activity

3.7 $\times$ **10**^{10} Bq/m^3 = **37** GBq/m^3

curie per kilogram Ci/kg

unit of specific activity (of radionuclide)

3.7 $\times$ **10**^{10} Bq/kg = **37** GBq/kg

cusec *cu*bic foot per *sec*ond (ft^3/s)

CV F abbr. for *cheval vapeur* = metric horsepower (—)

cwt hundredweight

cycle per second c/s
unit of frequency
1 hertz (q.v.)

D

d day; deci

D depr. symbol for deca (da) and dioptre (δ, dpt); darcy; debye

da deca

dag decagram

dalton Note: name sometimes used for the atomic mass unit (q.v.)

daraf (—)
Note: obsol. name used in US for reciprocal farad (F^{-1})

darcy (D)
unit of permeability (of porous material)
$9.869\,23 \times 10^{-1}$ μm^2

day d
jour
Tag
oSI unit of time
$\mathbf{8.64 \times 10^4}$ s
$\mathbf{1.44 \times 10^3}$ min
$\mathbf{2.4 \times 10}$ h
Note: symbol j formerly used in France

dB decibel

Note: sometimes accompanied by one or more letters to indicate special usage, e.g. dB(A) for 'A-weighted dB'. It is preferred to add the distinguishing term to the name, and the distinguishing letter as a subscript to the symbol, of the relevant quantity. For example, A-weighted sound pressure level $L_{pA}=80$ dB rather than sound pressure level $L_p=80$ dB (A).

debye (D)

obsol. unit of electric dipole moment

$3.335\,64\times10^{-30}$ C·m $=\mathbf{10^{-18}}$ Fr·cm

deca da

SI prefix denoting $\times 10$. It should be avoided as far as possible and used only where well established in practice. Examples: decagram (dag), decajoule (daJ), decalumen (dalm), decanewton (daN).

deci d

SI prefix denoting $\times 10^{-1}$. It should be avoided as far as possible and used only where well established in practice. Examples: decibel (dB), decigram (dg), decilitre (dl, dL) decimetre (dm).

decibel dB

décibel

Dezibel

unit of dimensionless quantities: amplitude level difference, power level difference, sound power level, sound pressure level, sound reduction index and sound intensity level

$n=k\lg(Q_1/Q_2)$

where n=number of decibels, Q_1 and Q_2=quantities of the same kind, $k=10$ or 20 depending on the quantities. For example, for power level difference and sound power level $k=10$, for amplitude level difference and sound pressure level $k=20$.

1 dB=0.115 129 Np [=(ln 10)/2)]

Note: see dB

decimilligrade . . . .^{cc}
depr. unit of plane angle
10^{-4} grade (q.v.)
Note: should be called one hundredth of a centigrade; also called centesimal second (q.v.)

deg E and F obsol. abbreviation for degree when used as a unit of temperature interval or difference. If unspecified by addition of another letter it refers to degree Kelvin (sometimes abbreviated degK) or degree Celsius (sometimes abbreviated degC), as degK=degC. See: temperature difference

degC obsol. abbr. for degree Celsius (q.v.) when used as a unit of temperature interval

degF obsol. abbr. for degree Fahrenheit (q.v.) when used as a unit of temperature interval

degK obsol. abbr. for degree Kelvin (see: kelvin) when used as a unit of temperature interval

degR obsol. abbr. for degree Rankine (q.v.) when used as a unit of temperature interval

degree . . . .°
degré
Grad; Altgrad
oSI unit of plane angle
$1.745\,33 \times 10^{-2}$ rad (=π/180)
6.0 ×**10** ′(minute)
3.6 ×**10^3** ″(second)
1.111 11 ^{g}(grade) (=400/360)
$1.111\,11 \times 10^{-2}$ ∟(right angle)
Note: it can also be subdivided decimally so that e.g. 4° 7′ 30″ is written as 4.125°. The degree with its decimal subdivision is recommended for use when radian is not suitable

degree (deg)
degré (deg)
Grad (grd)
depr. unit of temperature interval
Note: see deg

degree absolute
obsol. name for kelvin (q.v.)

degree Celsius °C
degré Celsius
Grad Celsius
SI unit of Celsius temperature and temperature interval
For temperatures:
$x°\,C=(x+273.15)K=(1.8x+32)\,°F$
$=(1.8x+491.67)\,°R$
For temperature intervals:
1°C = **1** K = **1.8** °F or °R
Note: cf. deg

degree centigrade depr. name for degree Celsius (q.v.)

degree Fahrenheit °F
degré Fahrenheit
Grad Fahrenheit
unit of Fahrenheit temperature and temperature interval
For temperatures:
$x\,°F=\frac{5}{9}\,(x+459.67)\,K=\frac{5}{9}\,(x-32)\,°C$
$=(x+459.67)\,°R$
For temperature intervals:
1°F = $\frac{5}{9}$ K or °C = **1** °R

degree Kelvin °K
former SI unit of thermodynamic temperature
Note: the name of this unit was changed at the 13th CGPM, 1967, to kelvin (q.v.) and its symbol to K. Also cf. deg

degree per second °/s
unit of angular velocity
$1.745\,33\times10^{-2}$ rad/s
1.047 20 rad/min

degree per second squared °/s²
unit of angular acceleration
$1.745\,33\times10^{-2}$ rad/s²

degree Rankine °R
degré Rankine
Grad Rankine
unit of Rankine temperature and temperature interval
For temperatures:
$x\,°\mathrm{R} = \frac{5}{9}x\,\mathrm{K} = \frac{5}{9}(x-491.67)\,°\mathrm{C} = (x-459.67)\,°\mathrm{F}$
For temperature intervals:
$1°\mathrm{R} = \frac{5}{9}\,\mathrm{K}$ or $°\mathrm{C} = \mathbf{1}°\mathrm{F}$

degree Réaumur (°R)
obsol. and depr. unit of temperature
Both for temperatures and temperature intervals:
$x\,°\mathrm{R}$ (degree Réaumur) $= 0.8x\,°\mathrm{C}$
Note: °R is used in this dictionary for degree Rankine (q.v.)

denier (den)
obsol. unit of linear density
$1\ \mathrm{g/9\ km} = 1.111\,11 \times 10^{-7}\ \mathrm{kg/m}$
Note: used for textile filaments

Dezitonne see: quintal

dg decigram; degree

dioptre (δ, dpt)
dioptrie
Dioptrie
unit of power of a lens
$\mathbf{1}\ \mathrm{m}^{-1}$
Note: US spelling is diopter

dk depr. symbol for deca (da)

dkg depr. symbol for decagram (dag)

dl, dL decilitre

dm decimetre

dollar (—)
unit of dimensionless quantity : reactivity
Equal to the effective delayed neutron fraction
$\mathbf{10^2}$ cent

dpt; dptr dioptre

dr dram (avoirdupois)

dr.ap.; dr ap apothecaries' drachm (or dram)

drachm, apothecaries' (—)
UK unit of mass
$3.887\,93 \times 10^{-3}$ kg
Note: the same as dram (apothecaries'); see 'apothecaries' units' and Table 10; cf. fluid drachm

dram, apothecaries' (dr ap)
US unit of mass
=drachm (apothecaries') (q.v.)
Note: cf. fluid dram

dram, avoirdupois (dr)
UK and US unit of mass
$1.771\,85 \times 10^{-3}$ kg
$\mathbf{2.734\,375 \times 10}$ gr (=7000/256)
M/S: Table 7

drex (—)
US and Canadian unit of linear density
1 g/**10** km = $\mathbf{10^{-1}}$ tex
Note: used for textile filaments

dry barrel (bbl)
US unit of volume (capacity) for dry measure
$1.156\,27 \times 10^{-1}$ m^3
$1.156\,27 \times 10^2$ dm^3 or litre
$\mathbf{7.056 \times 10^3}$ in^3
Note: this is the standard barrel used for fruits, vegetables and dry commodities

dry pint (dry pt)
US unit of volume (capacity) for dry measure
$5.506\,10 \times 10^{-4}$ m³
$5.506\,10 \times 10^{-1}$ dm³ or litre
$9.689\,39 \times 10^{-1}$ UKpt
M/S: Table 6

dry quart (US) (dry qt)
US unit of volume (capacity) for dry measure
1.101 22 dm^3
M/S: Table 6

dt G abbr. for *Dezitonne* = quintal (q.v.)

dwt pennyweight

dyn dyne

dyne dyn
dyne
Dyn
CGS unit of force
$= g \cdot cm/s^2$
10^{-5} N

dyne centimetre dyn·cm
CGS unit of moment of force
10^{-7} N·m = **1** erg (q.v.)

dyne centimetre per biot dyn·cm/Bi
CGSB unit of magnetic flux
10^{-8} Wb

dyne centimetre per second dyn·cm/s
CGS unit of moment of momentum
10^{-7} $kg \cdot m^2/s$ = **1** $g \cdot cm^2/s$ = erg second (q.v.)

dyne per biot centimetre dyn/(Bi·cm)
CGSB unit of magnetic flux density and magnetic polarization
For magnetic flux density:
10^{-4} T
For magnetic polarization;
$1.256\,64 \times 10^{-3}$ T

dyne per biot squared dyn/Bi^2
CGSB unit of permeability
$1.256\,64 \times 10^{-6}$ H/m

dyne per centimetre dyn/cm
CGS unit of surface tension
= erg/cm^2
10^{-3} N/m

dyne per cubic centimetre dyn/cm^3
CGS unit of specific weight
10 N/m^3

dyne per franklin dyn/Fr
CGSF unit of electric field strength
$2.997\,92 \times 10^4$ V/m

dyne per square centimetre dyn/cm^2
CGS unit of pressure
10^{-1} Pa = **10^{-6}** bar = **1** μbar
Note: also called barye

dyne second dyn·s
CGS unit of momentum
10^{-5} kg·m/s = **1** g·cm/s

dyne second per centimetre dyn·s/cm
CGS unit of mechanical impedance
10^{-3} N·s/m

dyne second per centimetre cubed dyn·s/cm^3
CGS unit of specific acoustic impedance
10 Pa·s/m

dyne second per centimetre to the fifth power dyn·s/cm^5
CGS unit of acoustic impedance
10^5 Pa·s/m^3

dyne second per square centimetre dyn·s/cm^2
= poise (q.v.)

dz G abbr. for *Doppelzentner* = quintal (q.v.)

E

E exa

electronvolt eV
électronvolt
Elektronvolt; Elektronenvolt
oSI unit of energy
$=1\ e \times 1$ V [e = elementary charge (q.v.)]
$1.602\ 189\ 2 \times 10^{-19}$ J ($C \times V = J$)

electronvolt per metre eV/m
oSI unit of linear stopping power and linear energy transfer
$1.602\ 19 \times 10^{-19}$ J/m

electronvolt per square metre eV/m^2
oSI unit of energy fluence
$1.602\ 19 \times 10^{-19}\ J/m^2$

electronvolt per square metre second $eV/(m^2 \cdot s)$
oSI unit of energy fluence rate
$1.602\ 19 \times 10^{-19}\ W/m^2$

electronvolt square metre $eV \cdot m^2$
oSI unit of atomic stopping power
$1.602\ 19 \times 10^{-19}\ J \cdot m^2$

electronvolt square metre per kilogram $eV \cdot m^2/kg$
oSI unit of mass stopping power
$1.602\ 19 \times 10^{-19}\ J \cdot m^2/kg$

emE G abbr. for *elektromagnetische Einheit* = electromagnetic unit

emu; e.m.u. abbr. used for CGSm units (e.g. emu of current, emu of permittivity). See: under ab- (e.g. abampere) or under CGSm unit (e.g. CGSm unit of permittivity)

engineer's chain (—)
unit of length
3.048 × **10** m
1.0 × **10**^{2} ft

erg erg
erg
Erg
CGS unit of work (energy)
10^{-7} joule (q.v.) = **1** dyn·cm

erg per biot (erg/Bi)
CGSB unit of magnetic flux
10^{-8} Wb = **10** nWb = **1** Mx

erg per biot squared (erg/Bi^{2})
CGSB unit of self inductance and mutual inductance
10^{-9} H

erg per centimetre erg/cm
CGS unit of linear stopping power and linear energy transfer
10^{-5} J/m
1 dyn

erg per cubic centimetre erg/cm^{3}
CGS unit of energy density and calorific value (volume basis)
10^{-1} J/m^{3}

erg per cubic centimetre degree Celsius erg/(cm^{3}·°C)
CGS unit of heat capacity per unit volume
10^{-1} J/(m^{3}·K)

erg per cubic centimetre second erg/(cm^{3}·s)
CGS unit of heat release rate
10^{-1} W/m^{3}

erg per centimetre second degree Celsius erg/(cm·s·°C)
CGS unit of thermal conductivity
10^{-5} W/(m·K)

erg per degree Celsius erg/°C
=erg per kelvin (q.v.)

erg per franklin (erg/Fr)
CGSF unit of electric potential
$2.997\,92 \times 10^{2}$ V

erg per gram erg/g
CGS unit of specific energy and kerma
$\mathbf{10^{-4}}$ J/kg or Gy

erg per gram degree Celsius erg/(g·°C)
CGS unit of specific heat capacity
$\mathbf{10^{-4}}$ J/(kg·K)

erg per gram second erg/(g·s)
CGS unit of absorbed dose rate and kerma rate
$\mathbf{10^{-4}}$ W/kg or Gy/s

erg per kelvin erg/K
CGS unit of heat capacity and entropy
$\mathbf{10^{-7}}$ J/K

erg per mole degree Celsius erg/(mol·°C)
CGS unit of molar gas constant
$\mathbf{10^{-7}}$ J/(mol·K)

erg per second erg/s
CGS unit of power and sound energy flux
$\mathbf{10^{-7}}$ W

erg per second steradian erg/(s·sr)
CGS unit of radiant intensity
$\mathbf{10^{-7}}$ W/sr

erg per second steradian square centimetre erg/(s·sr·cm^{2})
CGS unit of radiance
$\mathbf{10^{-3}}$ W/(sr·m^{2})

erg per square centimetre erg/cm^{2}
=dyne per centimetre (q.v.)

erg per square centimetre second $erg/(cm^2 \cdot s)$
CGS unit of energy fluence rate
10^{-3} W/m^2

erg per square centimetre second degree Celsius $erg/(cm^2 \cdot s \cdot °C)$
CGS unit of coefficient of heat transfer
10^{-3} $W/(m^2 \cdot s \cdot K)$

erg per square centimetre second kelvin to the fourth power $erg/(cm^2 \cdot s \cdot K^4)$
CGS unit of Stefan-Boltzmann constant
10^{-3} $W/(m^2 \cdot K^4)$

erg second erg·s
CGS unit of Planck constant
=dyn·cm·s
10^{-7} J·s or N·m·s

erg square centimetre $erg \cdot cm^2$
CGS unit of atomic stopping power
10^{-11} $J \cdot m^2$

erg square centimetre per gram $erg \cdot cm^2/g$
CGS unit of mass stopping power
10^{-8} $J \cdot m^2/kg$

erg square centimetre per second $erg \cdot cm^2/s$
CGS unit of first radiation constant
10^{-11} $W \cdot m^2$

esE G abbr. for *elektrostatische Einheit*=electrostatic unit (esu)

esu; e.s.u. abbr. used for CGSe units (e.g. esu of current, esu of permittivity). See: under stat- (e.g. statampere) or under CGSe unit (e.g. CGSe unit of permittivity)

eV electronvolt

exa E
SI prefix denoting $\times 10^{18}$. Examples: exahertz (EHz), exajoule (EJ), exaohm (EΩ)

F

f femto

F farad; femta

°F degree Fahrenheit

farad F
farad
Farad
SI unit of capacitance
$= m^{-2}\,kg^{-1}\,s^4\,A^2 = C/V = A{\cdot}s/V = s/\Omega$

farad per metre F/m
SI unit of permittivity
$= m^{-3}\,kg^{-1}\,s^4\,A^2 = C/(V{\cdot}m)$

farad square metre $F{\cdot}m^2$
= coulomb metre squared per volt (q.v.)

fathom (—)
unit of length
1.828 8 m = **2** yd
Note: for marine use

fbm board foot

fc foot candle

femta F
depr. prefix denoting $\times 10^{15}$

femto f
SI prefix denoting $\times 10^{-15}$. Example: femtoampere (fA), femtometre (fm), femtovolt (fV)

fermi (—)
unit of length
$\mathbf{10^{-15}}$ m = **1** fm (femtometre)
Note: used for nuclear distances

fg frigorie

fl dr fluid drachm (UK); fluid dram (US)

fl oz fluid ounce

fluid . . . – see also: liquid . . .

fluid drachm (UK) (UK fl dr)
UK unit of volume (capacity)
$3.551\,63 \times 10^{-6}$ m^3
M/S: Table 4

fluid dram (US) (US fl dr)
US obsol. unit of volume (capacity) for liquid measure
$3.696\,69 \times 10^{-6}$ m^3
M/S: Table 5

fluid ounce (UK) (UK fl oz)
UK unit of volume (capacity)
$2.841\,31 \times 10^{-5}$ m^3
$2.841\,31 \times 10$ cm^3
$1.733\,87$ in^3
$9.607\,60 \times 10^{-1}$ US fl oz
M/S: Table 4

fluid ounce (US) see: liquid ounce

fm fathom (—)

foot ft
UK and US unit, and FPS and ft-lbf-s base unit of length
$\mathbf{3.048 \times 10^{-1}}$ m
M/S: Table 1
Note: cf. survey foot

foot, board (—)
unit of volume
144 in^3 = $2.359\,74 \times 10^3$ m^3
Note: used in UK for timber

foot cubed ft^3
unit of modulus of section
Note: the same as cubic foot (q.v.)

foot hour degree Fahrenheit per British thermal unit
ft·h·°F/Btu
unit of thermal resistivity
$5.777\,89 \times 10^{-1}$ m·K/W
$2.419\,09 \times 10^{2}$ cm·s·°C/cal_{IT}
$6.719\,69 \times 10^{-1}$ m·h·°F/$kcal_{IT}$

foot of water (conventional) (ftH_2O)
unit of pressure
$2.989\,07 \times 10^{3}$ Pa

foot per minute ft/min
unit of velocity
$\mathbf{5.08} \times \mathbf{10^{-3}}$ m/s
$\mathbf{3.048} \times \mathbf{10^{-1}}$ m/min

foot per pound ft/lb
FPS unit of specific length
$6.719\,69 \times 10^{-1}$ m/kg

foot per second ft/s
FPS unit of velocity
$\mathbf{3.048} \times \mathbf{10^{-1}}$ m/s
1.097 28 km/h

foot per second squared ft/s^2
FPS unit of acceleration
$\mathbf{3.048} \times \mathbf{10^{-1}}$ m/s^2

foot pound ft·lb
this unit and symbol are often incorrectly used for foot pound-force (q.v.)

foot poundal ft·pdl
= poundal foot (q.v.)

foot poundal per second
FPS unit of power
$4.214\,01 \times 10^{-2}$ W

foot pound-force ft·lbf
ft-lbf-s unit of work
1.355 82 J
$3.766\ 16 \times 10^{-7}$ kW·h

foot pound-force per pound ft·lbf/lb
unit of specific internal energy and specific latent heat
2.989 07 J/kg
$7.139\ 26 \times 10^{-4}$ $kcal_{IT}$/kg
3.048 $\times \mathbf{10^{-1}}$ kgf·m/kg or kp·m/kg
$1.285\ 07 \times 10^{-3}$ Btu/lb

foot pound-force per pound degree Fahrenheit
ft·lbf/(lb·°F)
unit of specific heat capacity
5.380 32 J/(kg·K)
$1.285\ 07 \times 10^{-3}$ $kcal_{IT}$/(kg·°C) or Btu/(lb·°F)
5.486 4 $\times \mathbf{10^{-1}}$ kgf·m/(kg·°C) or kp·m/(kg·°C)

foot pound-force per second ft·lbf/s
ft-lbf-s unit of power
1.355 82 W

foot squared per hour ft^2/h
unit of kinematic viscosity
2.580 64 $\times \mathbf{10^{-5}}$ m^2/s
2.580 64 $\times$ **10** cSt
$2.777\ 78 \times 10^{-4}$ ft^2/s

foot squared per second ft^2/s
FPS unit of kinematic viscosity
$9.290\ 30 \times 10^{-2}$ m^2/s
$9.290\ 30 \times 10^{4}$ cSt
3.6 $\times \mathbf{10^{3}}$ ft^2/h

foot to the fourth power ft^4
FPS unit of second moment of area
$8.630\ 97 \times 10^{-3}$ m^4

foot-candle (fc)
unit of illuminance
=lumen per square foot (q.v.)

foot-lambert (ft·La)
unit of luminance
3.426 26 cd/m²

fr frigorie

Fr franklin; French scale (see: charrière)

franklin (Fr)
CGSF unit of electric charge and electric flux
For electric change:
$3.335\,64 \times 10^{-10}$ C
For electric flux:
$2.654\,42 \times 10^{-11}$ C

franklin centimetre (Fr·cm)
CGSF unit of electric dipole moment
$3.335\,64 \times 10^{-12}$ C·m

franklin per second (Fr/s)
CGSF unit of electric current
$3.335\,64 \times 10^{-10}$ A

franklin per square centimetre (Fr/cm^2)
CGSF unit of electric polarization and electric flux density
For polarization:
$3.335\,64 \times 10^{-6}$ C/m^2
For electric flux density
$2.654\,42 \times 10^{-7}$ C/m^2

franklin squared per erg (Fr^2/erg)
CGSF unit of capacitance
$1.112\,65 \times 10^{-12}$ F

franklin squared per erg centimetre ($Fr^2/(erg \cdot cm)$)
CGSF unit of permittivity
$8.854\,19 \times 10^{-12}$ F/m

freight ton (—)
tonneau d'encombrement
Raumtonne
unit of ship cargo
40 $ft^3 = 1.132\,674\ m^3$
Note: also called shipping ton or measurement ton

frigorie fg
unit of heat (for refrigeration)
$4.185\,5 \times 10^3$ J
1.0 $kcal_{15}$
Note: sometimes its negative value is used

frigorie per hour fg/h
unit of refrigerating capacity
1.162 64 W

ft foot

ftH_2O conventional foot of water

ft L; ft La foot lambert

furlong (—)
US and UK unit of length
$\mathbf{2.011\,68 \times 10^2}$ m
M/S:Table 1
Note: obsol. but still used in horse-racing

G

. . . .^{g} gon; grad

g gram

g* gram-weight

G giga; gauss

g; G unit of acceleration equal to standard acceleration of free fall (q.v.)

gal Gal
gal
Gal
CGS unit of (linear) acceleration
$\mathbf{10^{-2}}$ m/s^2 = **1** cm/s^2
Note: formerly called galileo

gal gallon

Gal gal

galileo see: gal

gallon (UK) (UKgal)
UK unit of volume ('capacity')
$\mathbf{4.546\,09 \times 10^{-3}}$ m^3
4.546 09 dm^3 or l or L
$1.605\,44 \times 10^{-1}$ ft^3
$2.774\,20 \times 10^{2}$ in^3
1.200 95 USgal
M/S: Table 4
Note: also called imperial gallon. It is the fundamental UK unit of capacity defined by the WMA, 1963, and redefined by UMR, 1976, as **4.546 09** dm^3. The same value is accepted in Canada and Australia

gallon (UK) per hour (UKgal/h)
UK unit of volume flow rate
$1.262\,80 \times 10^{-6}$ m^3/s
$4.546\,09 \times 10^{-3}$ m^3/h

gallon (UK) per mile (UKgal/mile)
UK unit of fuel consumption
2.824 81 litres/km

gallon (UK) per minute (UKgal/min)
UK unit of volume flow rate
$7.576\,82 \times 10^{-5}$ m^3/s
$2.727\,65 \times 10^{-1}$ m^3/h

gallon (UK) per pound (UKgal/lb)
UK unit of specific volume
$1.002\,24 \times 10^{-2}$ m³/kg

gallon (UK) per second (UKgal/s)
UK unit of volume flow rate
$4.546\,09 \times 10^{-3}$ m³/s
$1.636\,59 \times 10$ m³/h

gallon (US) (USgal)
US unit of volume (capacity) for liquid measure

3.785 411 784	$\times 10^{-3}$	m^3	
3.785 41		dm^3 or l or L	
1.336 81	$\times 10^{-1}$	ft^3	
2.31	$\times$ **10**^{2}	in^3	(=definition)
$8.326\,74 \times 10$	$\times 10^{-1}$	UKgal	

M/S: Table 5 and barrel (q.v.)

gallon (US) per hour (USgal/h)
US unit of volume flow rate
$1.051\,50 \times 10^{-6}$ m³/s
$3.785\,41 \times 10^{-3}$ m³/h

gallon (US) per mile (USgal/mile)
US unit of fuel consumption
2.352 15 litres/km

gallon (US) per minute (USgal/min)
US unit of volume flow rate
$6.309\,02 \times 10^{-5}$ m³/s

gallon (US) per pound (USgal/lb)
US unit of specific volume
$8.345\,40 \times 10^{-3}$ m³/kg

gallon (US) per second (USgal/s)
US unit of volume flow rate
$3.785\,41 \times 10^{-3}$ m³/s

gamma (γ)
(a) unit of mass
10^{-9} kg = **1** μg
(b) unit of magnetic flux density
10^{-9} T = 1 nT

gauss (Gs; G)
CGSm unit of magnetic flux density
10^{-4} T

Gaussian CGS unit of . . . see: CGSe unit of . . ., CGSm unit of . . ., ab- and stat-

Gb gilbert

gee pound see: slug

gf gram-force

gi gill (US)

giga G
SI prefix denoting $\times 10^9$. Examples: gigabecquerel (GBq), gigacalorie (Gcal), gigaelectronvolt (GeV), gigahertz (GHz), gigajoule (GJ), giganewton (GN), gigaohm (GΩ), gigapascal (GPa), gigapond (Gp), gigawatt (GW)

gilbert (Gb)
CGSm unit of magnetomotive force
$7.957\,75 \times 10^{-1}$ A [$= 10/(4\pi)$]

gilbert per centimetre (Gb/cm)
CGSm unit of magnetic field strength
Note: corresponds to oersted (q.v.)

gilbert per maxwell (Gb/Mx)
CGSm unit of reluctance
$7.957\,75 \times 10^{7}$ H^{-1}

gill (UK) (—)
UK unit of volume (capacity)
$1.420\,65 \times 10^{-4}$ m^3
M/S: Table 4

gill (US) (gi)
US unit of volume (capacity) for liquid measure
$1.182\,94 \times 10^{-4}$ m^3
M/S: Table 5

gm gram (g)

gon $\ldots^{g}$
unit of plane angle
Note: see grade

gpm gallon per minute (gal/min)

gps gallon per second (gal/s)

gr grain; grade ($\ldots^{g}$); gram (g)

g·rad gram-rad

grade $\ldots^{g}$
grade
Gon; Neugrad
unit of plane angle
$1.570\,80 \times 10^{-2}$ rad (=π/200)
0.9° (degree) = **54**′ (minute) = **3240**″ (second)
0.01∟ (right angle)
Note: also called gon; the grade is subdivided decimally

grade per second ^{g}/s
unit of angular velocity
$1.570\,80 \times 10^{-2}$ rad/s

grade per second squared ^{g}/s^2
unit of angular acceleration
$1.570\,80 \times 10^{-2}$ rad/s^2

grain gr
UK and US unit of mass
$6.479\,891 \times 10^{-5}$ kg
$6.479\,891 \times 10$ mg
$1.428\,57 \times 10^{-4}$ lb (=1/7000)
M/S: Tables 9 & 10
Note: no symbol is used in US; grain is an important unit common to and linking the avoirdupois, apothecaries' and troy systems

grain per cubic foot gr/ft^3
unit of (mass) density and concentration
$2.288\,35 \times 10^{-3}$ kg/m^3

grain per UK gallon gr/UKgal
UK unit of (mass) density and concentration
$1.425\,38 \times 10^{-2}$ kg/m^3

grain per US gallon gr/USgal
US unit of (mass) density and concentration
$1.711\,81 \times 10^{-2}$ kg/m^3

gram g
gramme
Gramm
mSI unit and CGS base unit of mass
10^{-3} kg
Note: formerly spelled gramme

gram centimetre per second g·cm/s
CGS unit of momentum
10^{-5} kg·m/s = **1** dyn·s

gram centimetre per second squared g·cm/s^2
=dyne (q.v.)

gram centimetre squared g·cm^2
CGS unit of moment of inertia
10^{-7} kg·m^2

gram centimetre squared per second $g{\cdot}cm^2/s$
CGS unit of moment of momentum
10^{-7} $kg{\cdot}m^2/s$ = **1** dyn·cm/s

gram per cubic centimetre g/cm^3
CGS unit of (mass) density
10^3 kg/m^3
1 kg/dm^3 (q.v.)

gram per litre g/l, g/L
unit of (mass) density
1 kg/m^3

gram per millilitre g/ml, g/mL
unit of (mass) density
10^3 kg/m^3
1 kg/dm^3 (q.v.)

gram per square metre g/m^2
mSI unit of surface density and grammage of paper or paperboard
10^{-3} kilogram per square metre (q.v.)

gram per square metre day $g/(m^2{\cdot}d)$
oSI unit of water vapour transmission rate

gram-atom (—)
obsol. unit of mass of an element
$= n$ grams,where $n =$ relative atomic mass of that element

gram-calorie (—)
obsol. name for calorie (q.v.)

gram-force gf
m-kgf-s unit of force
$9.806\,65 \times 10^{-3}$ N

gram-molecule (gmol)
obsol. unit of mass of a compound
$= n$ grams, where $n =$ relative molecular mass of that compound

gram-rad (g·rad)
unit of integral absorbed dose
10^{-5} J = **10** μJ = **10^2** erg

gram-weight (g*, g(wt))
obsol. name for gram-force

gray Gy
SI unit of absorbed dose, specific energy imparted and kerma
$m^2\,s^{-2}$
1 J/kg = **10^2** rad

gray per second Gy/s
SI unit of absorbed dose rate and kerma rate
$m^2\,s^{-3}$
1 W/kg = **10^2** rad/s

grd G obsol. abbr. for *Grad* = degree

Gs gauss

gsm gram per square metre (g/m^2)

Gy gray

H

h hecto; hour

H henry

ha hectare

hand (—)
unit of length
1.016 × **10^{-1}** m = **4** in
Note: obsol. but still used for measuring the height of horses

hebdo (—)
depr. prefix denoting $\times 10^7$

hectare ha
hectare
Hektar
unit of area
$\mathbf{10^4}$ m^2 = $\mathbf{10^2}$ a = $\mathbf{10^{-2}}$ km^2
$3.861\,02 \times 10^{-3}$ mile^2
2.471 05 acre
$1.195\,99 \times 10^4$ yd^2
Note: used for agrarian measurements only

hectare-millimetre (ha·mm)
unit of volume
10 m^3
Note: used very rarely

hecto h
SI prefix denoting $\times 10^2$. It should be avoided as far as possible and used only where well established in practice. Examples: hectare (ha), hectolitre (hl), hectopièze (hpz)

hectolitre hl
hectolitre
Hektoliter
unit of volume (capacity)
$\mathbf{10^{-1}}$ m^3 = $\mathbf{10^2}$ dm^3 or litre
Note: used e.g. in brewing industry

hectopièze (hpz)
unit of pressure
$\mathbf{10^5}$ Pa = **1** bar
Note: formerly used in France

Hefner candle
obsol. unit of luminous intensity
0.903 cd
Note: until 1942 used widely in Germany under the name *Hefner-Kerze,* symbol HK

henry H

henry
Henry
SI unit of self inductance, mutual inductance and permeance
$= m^2\ kg\ s^{-2}\ A^{-2} = V{\cdot}s/A = Wb/A = \Omega{\cdot}s = J/A^2$

henry per metre H/m

SI unit of permeability
$= m\ kg\ s^{-2}\ A^{-2} = N/A^2$

hertz Hz

hertz
Hertz
SI unit of frequency
1 s^{-1} = **1** c/s

HK see: Hefner candle

hl hectolitre

horsepower (metric) (—)

cheval vapeur (ch)
Pferdestärke (PS)
unit of power
$7.354\,99 \times 10^2$ W (= **75** × **9.806 65**)
7.5 × **10** kgf·m/s or kp·m/s
$9.863\,20 \times 10^{-1}$ hp

horsepower (hp)

UK and US unit of power
$7.457\,00 \times 10^2$ W
5.5 × **10**^{2} ft·lbf/s
1.013 87 metric horsepower

horsepower hour (metric) (—)

cheval-heure (ch h)
Pferdestärkenstunde (PSh)
unit of work
$2.647\,80 \times 10^6$ J
$7.354\,99 \times 10^{-1}$ kW·h

horsepower hour hp·h
UK and US unit of energy
$2.684\,52 \times 10^{6}$ J
$7.457\,00 \times 10^{-1}$ kW·h

hour h
heure
Stunde
oSI unit of time
3600 s = **60** min
$4.166\,67 \times 10^{-2}$ d

hp horsepower

hp·h horsepower hour

hpz hectopièze

hr depr. abbr. for hour (h)

hundredweight (cwt)
UK unit of mass
$5.080\,23 \times 10$ kg
1.12 short hundredweight
M/S: Table 7
Note: also called long hundredweight to distinguish it from short hundredweight (q.v.)

hundredweight see also: short hundredweight

hyl (—)
m-kgf-s unit of mass
= gf·s²/m or p·s²/m
$\mathbf{9.806\,65 \times 10^{-3}}$ kg
Note: rarely used; often confused with kilohyl (q.v.)

Hz hertz

I

IC international candle

IK G abbr. for *Internationale Kerze* = international candle (q.v.)

i.; imp. imperial

imperial (imp.)
Adjective indicating that a particular unit is one of the so-called imperial units (q.v.)

imperial units Imperial units lawful for use for trade in the United Kingdom under the Weight and Measures Act, 1963 and related Acts and Regulations are the following:

(a) units of length: mile, furlong, chain, yard, foot, inch

(b) units of area: square mile, acre, rood, square yard, square foot, square inch

(c) units of volume: cubic yard, cubic foot, cubic inch

(d) units of 'capacity': gallon, quart, pint, gill, fluid ounce

(e) units of mass or 'weight': ton, hundredweight, cental, quarter, stone, pound, ounce, dram, grain

Note: see 'apothecaries' units' and 'troy units'. Bushel, peck, fluid drachm and minim are no longer lawful

in inch

inHg conventional inch of mercury

inH$_2$O conventional inch of water

inch in
(*pouce*)
(*Zoll*)
UK and US unit of length
2.54 $\times$ **10**$^{-2}$ m = **25.4** mm
2.777 78 $\times$ 10^{-2} yd
8.333 33 $\times$ 10^{-2} ft
M/S: Table 1

inch cubed in^3
unit of modulus of section
Note: the same as cubic inch (q.v.)

inch of mercury (conventional) (inHg)
unit of pressure
3.386 39 $\times$ 10^3 Pa

inch of water (conventional) (inH_2O)
unit of pressure
2.490 89 $\times$ 10^2 Pa

inch per minute in/min
unit of velocity
4.233 33 $\times$ 10^{-4} m/s
2.54 $\times$ **10**$^{-2}$ m/min

inch per second in/s
unit of velocity
2.54 $\times$ **10**$^{-2}$ m/s

inch squared per hour in^2/h
unit of kinematic viscosity
1.792 11 $\times$ 10^{-7} m^2/s
1.792 11 $\times$ 10^{-1} cSt
6.451 6 $\times$ **10**$^{-4}$ m^2/h
1.929 01 $\times$ 10^{-6} ft^2/s

inch squared per second in^2/s
unit of kinematic viscosity
6.451 6 $\times$ **10**$^{-4}$ m^2/s
6.451 6 $\times$ **10**^{2} cSt
6.944 44 $\times$ 10^{-3} ft^2/s

inch to the fourth power in^4
unit of second moment of area
$4.162\,31 \times 10^{-7}$ m^4

inhour (—)
unit of dimensionless quantity: reactivity. Equal to the increase in reactivity of a critical reactor which produces a reactor time constant of 1 hour. (Short for inverse hour)

INM abbr. sometimes used for international nautical mile

international ampere (A_{int})
obsol. unit of electric current
$9.998\,5 \times 10^{-1}$ A

international candle (IC)
obsol. unit of luminous intensity
1.02 cd

international coulomb (C_{int})
obsol. unit of electric charge
$9.998\,5 \times 10^{-1}$ C

international farad (F_{int})
obsol. unit of capacitance
$9.995\,1 \times 10^{-1}$ F

international henry (H_{int})
obsol. unit of inductance and permeance
1.000 49 H

international joule (mean) (J_{int})
obsol. unit of work, energy and heat
1.000 19 J

international nautical mile see: nautical mile (international)

international ohm (Ω_{int})
obsol. unit of resistance
1.000 49 Ω

international siemens (S_{int})
obsol. unit of conductance
$9.995\,1 \times 10^{-1}$ S

international table calorie cal_{IT}
calorie I.T.
internationale Tafel-Kalorie
unit of heat
Note: in this dictionary referred to as calorie (I.T.)

international table kilocalorie $kcal_{IT}$
kilocalorie I.T.
Internationale Tafel-Kilokalorie
unit of heat
Note: in this dictionary referred to as kilocalorie (I.T.)

international tesla (T_{int})
obsol. unit of magnetic flux density
1.000 34 T

international volt (V_{int})
obsol. unit of electric potential
1.000 34 V

international watt (W_{int})
obsol. unit of power
1.000 19 W

international weber (Wb_{int})
obsol. unit of magnetic flux
1.000 34 Wb

ipm, ips inches per minute (in/min), inches per second (in/s)

ipr inches per revolution (in/r)

J

J joule

joule J
joule
Joule
SI unit of work, energy and heat
$= m^2\ kg\ s^{-2} = N{\cdot}m = W{\cdot}s = Pa{\cdot}m^3 = C{\cdot}V$
$6.241\ 46 \times 10^{18}$ eV
$2.777\ 78 \times 10^{-7}$ kW·h
$\mathbf{1.0 \times 10^7}$ erg or dyn·cm
$1.019\ 72 \times 10^{-1}$ kgf·m or kp·m
$2.388\ 46 \times 10^{-4}$ $kcal_{IT}$
$7.375\ 62 \times 10^{-1}$ ft·lbf
$9.478\ 17 \times 10^{-4}$ Btu
Note: also unit of enthalpy, exchange integral, level width and work function, and also exergy and anergy

joule per cubic metre J/m^3
SI unit of energy density
$= m^{-1}\ kg\ s^{-2} = W{\cdot}s/m^3 = N/m^2$
$\mathbf{1.0 \times 10}$ erg/cm^3
$2.388\ 46 \times 10^{-4}$ $kcal_{IT}/m^3$
$2.683\ 92 \times 10^{-5}$ Btu/ft^3
Note: also unit of calorific value (volume basis) and refrigerating capacity per unit volume

joule per degree Celsius J/°C
=joule per kelvin (q.v.)

joule per kelvin J/K
SI unit of heat capacity and entropy
$= m^2\ kg\ s^{-2}\ K^{-1} = J/°C$
$2.388\ 46 \times 10^{-4}$ $kcal_{IT}/°C$
Note: also unit of Boltzmann constant, Massieu function and Planck function

joule per kilogram J/kg

SI unit of specific energy and specific enthalpy

$m^2\,s^{-2}$ = W·s/kg = N·m/kg = **1** Gy = **1** Sv

$2.388\,46 \times 10^{-4}$ $kcal_{IT}/kg$ or cal_{IT}/g

1.0 $\times$ $\mathbf{10^4}$ erg/g

$6.241\,46 \times 10^{-21}$ eV/g

$1.019\,72 \times 10^{-1}$ kgf·m/kg or kp·m/kg

$4.299\,23 \times 10^{-4}$ Btu/lb

$3.345\,53 \times 10^{-1}$ ft·lbf/lb

Note: also unit of specific latent heat and calorific value (mass basis), and also specific exergy and anergy

joule per kilogram degree Celsius J/(kg·°C)

= joule per kilogram kelvin (q.v.)

joule per kilogram kelvin J/(kg·K)

SI unit of specific heat capacity and specific entropy

$= m^2\,s^{-2}\,K^{-1}$ = J/(kg·°C)

$2.388\,46 \times 10^{-4}$ $kcal_{IT}$/(kg·°C) or Btu/(lb·°F)

1.0 $\times$ $\mathbf{10^4}$ erg/(g·°C)

$1.019\,72 \times 10^{-1}$ kgf·m/(kg·°C) or kp·m/(kg·°C)

$1.858\,63 \times 10^{-1}$ ft·lbf/(lb·°F)

joule per kilogram second J/(kg·s)

= watt per kilogram (q.v.)

joule per metre J/m

SI unit of (total) linear stopping power and linear energy transfer

$= m\,kg\,s^{-2}$ = newton (q.v.)

$6.241\,46 \times 10^{18}$ eV/m

1.0 $\times$ $\mathbf{10^5}$ erg/cm or dyne

joule per metre to the fourth power J/m^4

SI unit of spectral concentration of radiant energy density (in terms of wavelength)

$= m^{-2}\,kg\,s^{-2} = N/m^3$ = Pa/m

joule per mole J/mol

SI unit of molar internal energy

$= m^2\,kg\,s^{-2}\,mol^{-1}$

Note: also unit of chemical potential and affinity

joule per mole degree Celsius J/(mol·°C)
=joule per mole kelvin (q.v.)

joule per mole kelvin J/(mol·K)
SI unit of molar heat capacity and molar entropy
$=m^2\ kg\ s^{-2}\ K^{-1}\ mol^{-1}$
Note: also unit of molar gas constant

joule per pound kelvin or degree Celsius J/(lb·K) or J/(lb·°C)
unit of specific heat capacity and specific entropy
2.204 62 J/(kg·K)

joule per second J/s
=watt (q.v.)

joule per square metre J/m^2
SI unit of energy fluence and radiant exposure
$=kg\ s^{-2}=N/m$
$6.241\ 46\times10^{18}\ eV/m^2$

joule per square metre second $J/(m^2{\cdot}s)$
=watt per square metre (q.v.)

joule per square metre second kelvin $J/(m^2{\cdot}s{\cdot}K)$
=watt per square metre kelvin (q.v.)

joule per tesla J/T
=ampere square metre (q.v.)

joule reciprocal hertz $J{\cdot}Hz^{-1}$
=joule second (q.v.)

joule reciprocal tesla $J{\cdot}T^{-1}$
=ampere square metre (q.v.)

joule second J·s
SI unit of Planck constant and action
$=m^2\ kg\ s^{-1}=N{\cdot}m{\cdot}s$
10^7 erg·s or dyn·cm·s

joule square metre J·m²
SI unit of (total) atomic stopping power
$= m^4$ kg s^{-2}
10^{11} erg·cm²

joule square metre per kilogram J·m²/kg
SI unit of (total) mass stopping power
$= m^4 s^{-2}$
10^8 erg cm²/g

Julian year (—)
unit of time
$3.155\,76 \times 10^7$ s
$5.259\,6 \times 10^5$ min
8.766×10^3 h
$3.652\,5 \times 10^2$ d

K

k kilo

K kelvin; kilobyte; kayser

°K degree Kelvin

karat see: carat

kayser (K)
obsol. unit of wavenumber
10^2 m^{-1} = **1** cm^{-1}

kc kilocycle (per second) = kilohertz (kHz)

kcal kilocalorie

$kcal_{15}$ kilocalorie 15 °C

$kcal_{IT}$ kilocalorie I.T.

$kcal_{th}$ thermochemical kilocalorie

kelvin K
kelvin
Kelvin
base SI unit of thermodynamic temperature and SI unit of temperature interval (difference) and other temperatures
Def: see Appendix 1
For temperatures:
$x\,\mathrm{K}=(x-273.15)\,°\mathrm{C}=(1.8x-459.67)\,°\mathrm{F}=1.8x\,°\mathrm{R}$
For temperature intervals:
$1\,\mathrm{K}=1\,°\mathrm{C}=1.8\,°\mathrm{F}$ or °R

kelvin per metre K/m
SI unit of temperature gradient
=°C/m

kelvin per watt K/W
SI unit of thermal resistance
$=\mathrm{m}^{-2}\,\mathrm{kg}^{-1}\,\mathrm{s}^{3}\,\mathrm{K}$

kg kilogram

kg* kilogram-weight

kgf kilogram-force

kgph kilogram per hour (kg/h)

kgpm kilogram per minute (kg/min)

kgps kilogram per second (kg/s)

kilo k
SI prefix denoting $\times 10^3$. Examples: kiloampere (kA), kilobar (kbar), kilobecquerel (kBq), kilocalorie (kcal), kilocoulomb (kC), kiloelectronvolt (keV), kilogram (kg), kilogray (kGy), kilohertz (kHz), kilojoule (kJ), kilolux (klx), kilometre (km), kilomole (kmol), kilonewton (kN), kiloohm (kΩ), kilopascal (kPa), kilopond (kp), kilosecond (ks), kilosiemens (kS), kilotex (ktex), kilovolt (kV), kilowat (kW), kiloweber (kWb).
Note: when referring to memory capacity the prefix (or its symbol) denotes *approximately* $\times 10^3$, usually $\times 1024$ $(=2^{10})$, e.g. kilobyte (kbyte, often Kb)$=2^{10}$ byte

kilocalorie (I.T.) $kcal_{IT}$ (or kcal)
obsol. unit of heat
4.1868 × **10**³ J = **10**³ cal_{IT}
Note: for other values see 'calorie'

kilocalorie (I.T.) metre per square metre hour kelvin or degree Celsius $kcal_{IT}\cdot m/(m^2\cdot h\cdot K)$ or $kcal_{IT}\cdot m/(m^2\cdot h\cdot °C)$
unit of thermal conductivity
1.163 W/(m·K)

kilocalorie (I.T.) per cubic metre $kcal_{IT}/m^3$
unit of calorific value (volume basis)
4.186 8 × **10**³ J/m³
1.123 70 × 10⁻¹ Btu/ft³

kilocalorie (I.T.) per cubic metre hour $kcal_{IT}/(m^3\cdot h)$
unit of heat release rate
1.163 W/m³
1.123 70 × 10⁻¹ Btu/(ft³·h)

kilocalorie (I.T.) per hour $kcal_{IT}/h$
unit of heat flow rate
1.163 W
3.968 32 Btu/h

kilocalorie (I.T.) per kelvin or degree Celsius
$kcal_{IT}/K$ or $kcal_{IT}/°C$
unit of heat capacity
4.186 8 × **10**³ J/K

kilocalorie (I.T.) per kilogram $kcal_{IT}/kg$
unit of specific internal energy and calorific value (mass basis)
4.186 8 × **10**³ J/kg
1.8 Btu/lb

kilocalorie (I.T.) per kilogram kelvin or degree Celsius
$kcal_{IT}/(kg\cdot K)$ or $kcal_{IT}/(kg\cdot °C)$
unit of specific heat capacity
4.186 8 × **10**³ J/(kg·K)
1 Btu/(lb·°R) or Btu/(lb·°F)

kilocalorie (I.T.) per metre hour kelvin or degree Celsius
$kcal_{IT}/(m \cdot h \cdot K)$ or $kcal_{IT}/(m \cdot h \cdot °C)$

unit of thermal conductivity
1.163 W/(m·K)
$6.719\,69 \times 10^{-1}$ Btu/(ft·h·°R) or Btu/(ft·h·°F)

kilocalorie (I.T.) per square metre hour $kcal_{IT}/(m^2 \cdot h)$

unit of density or heat flow rate
1.163 W/m^2
$3.686\,69 \times 10^{-1}$ $Btu/(ft^2 \cdot h)$

kilocalorie (I.T.) per square metre hour kelvin or degree Celsius $kcal_{IT}/(m^2 \cdot h \cdot K)$ or $kcal_{IT}/(m^2 \cdot h \cdot °C)$

unit of coefficient of heat transfer
1.163 $W/(m^2 \cdot K)$
$2.048\,16 \times 10^{-1}$ $Btu/(ft^2 \cdot h \cdot °R)$ or $Btu/(ft^2 \cdot h \cdot °F)$

kilogram kg

kilogramme
Kilogramm
base SI unit of mass
Def: see Appendix 1
$\mathbf{10^{-3}}$ t = $\mathbf{10^{3}}$ g
5.0 $\times$ $\mathbf{10^{3}}$ metric carat
2.204 62 lb
$3.527\,40 \times 10$ oz
$1.543\,24 \times 10^4$ gr
Note: also SI unit of (rest) mass of particles, mass defect and mass excess; formerly spelled kilogramme

kilogram metre kg·m

Note: this name and symbol are often used incorrectly for kilogram-force metre (q.v.)

kilogram metre per second kg·m/s

SI unit of momentum
= N·s
1.0 $\times$ $\mathbf{10^{5}}$ g·cm/s or dyn·s
7.233 01 lb·ft/s

kilogram metre per second squared $kg \cdot m/s^2$

= newton (q.v.)

kilogram metre squared kg·m^2
SI unit of moment of inertia
m^2 kg = J·s^2
1.0 × **10**^{7} g·cm^2
2.373 04 × 10 lb·ft^2
3.417 17 × 10^3 lb·in^2

kilogram metre squared per second kg·m^2/s
SI unit of moment of momentum
m^2 kg s^{-1} = N·m·s
1.0 × **10**^{7} g·cm^2/s
2.373 04 × 10 lb·ft^2/s

kilogram per cubic centimetre kg/cm^3
mSI unit of (mass) density
10^6 kg/m^3

kilogram per cubic decimetre kg/dm^3
mSI unit of (mass) density
10^3 kg/m^3 = **1** kg/l, kg/L or g/cm^3 or g/ml, g/mL or t/m^3 or Mg/m^3

kilogram per cubic metre kg/m^3
SI unit of (mass) density
1.0 × **10**$^{-3}$ g/cm^3
6.242 80 × 10^{-2} lb/ft^3
Note: also unit of mass concentration

kilogram per cubic metre pascal kg/(m^3·Pa)
SI unit of unitary mass density
= m^{-2} s^2

kilogram per hectare kg/ha
unit of surface density
10^{-4} kilogram per square metre (q.v.)

kilogram per hour kg/h
oSI unit of mass flow rate
2.777 78 × 10^{-4} kg/s
6.123 95 × 10^{-4} lb/s
2.204 62 lb/h

kilogram per litre kg/l or kg/L
oSI unit of (mass) density
=kilogram per cubic decimetre (q.v.)

kilogram per metre kg/m
SI unit of linear density
1.0 t/km
$6.719\ 69 \times 10^{-1}$ lb/ft
$5.599\ 74 \times 10^{-2}$ lb/in
2.015 91 lb/yd
Note: used for wires etc.

kilogram per metre second kg/(m·s)
=pascal second (q.v.)

kilogram per mole kg/mol
SI unit of molar mass
10^3 g/mol or kg/kmol

kilogram per pascal second metre kg/(Pa·s·m)
SI unit of water vapour permeance

kilogram per pascal second square metre kg/(Pa·s·m^2)
SI unit of water vapour permeability

kilogram per second kg/s
SI unit of mass flow rate
3.6 $\times$ **10^3** kg/h
2.204 62 lb/s
$7.936\ 64 \times 10^3$ lb/h

kilogram per square centimetre kg/cm^2
mSI unit of surface density
10^4 kg/m^2
Note: this name and symbol are often used incorrectly for kilogram-force per square centimetre (q.v.)

kilogram per square metre kg/m^2
SI unit of surface density
$2.048\ 16 \times 10^2$ lb/1000 ft^2
$8.921\ 79 \times 10^3$ lb/acre
$2.949\ 35 \times 10$ oz/yd^2
Note: used in agriculture and for sheet metal and plating. Also unit of mean mass range

kilogram-calorie kcal
obsol. name for kilocalorie (q.v.)

kilogram-force (kgf)
kilogramme-force (kgf)
Kilopond (kp)
m-kgf-s base unit of force
9.806 65 N
$7.093\,16 \times 10$ pdl
2.204 62 lbf

kilogram-force metre (kgf·m)
mètre-kilogramme-force
Kilopondmeter
m-kgf-s unit of moment of force and torque; work and energy
9.806 65 N·m or J
$2.724\,07 \times 10^{-6}$ kW·h
7.233 01 lbf·ft

kilogram-force metre per kilogram (kgf·m/kg)
m-kgf-s unit of specific internal energy and specific latent heat
9.806 65 J/kg

kilogram-force metre per kilogram degree Celsius (kgf·m/(kg·°C))
m-kgf-s unit of specific heat capacity
9.806 65 J/(kg·K)

kilogram-force metre per second (kgf·m/s)
m-kgf-s unit of power
9 806 65 W

kilogram-force metre second (kgf·m·s)
m-kgf-s unit of action
9.806 65 J·s or N·m·s

kilogram-force metre second squared (kgf·m·s^2)
m-kgf-s unit of moment of inertia
9.806 65 kg·m^2

kilogram-force per centimetre (kgf/cm)
m-kgf-s unit of surface tension
9.806 65 $\times 10^2$ N/m

kilogram-force per cubic metre (kgf/m^3)
m-kgf-s unit of specific weight
9.806 65 N/m^3

kilogram-force per metre (kgf/m)
m-kgf-s unit of surface tension
9.806 65 N/m

kilogram-force per metre second degree Celsius (kgf/(m·s·°C))
m-kgf-s unit of coefficient of heat transfer
9.806 65 W/(m^2·K)

kilogram-force per second degree Celsius (kgf/(s·°C))
m-kgf-s unit of thermal conductivity
9.806 65 W/(m·K)

kilogram-force per square centimetre (kgf/cm^2)
m-kgf-s unit of pressure
9.806 65 $\times$ **10^4** Pa
9.806 65 $\times$ **10^{-1}** bar
1.0 at
1.422 33 $\times$ 10 lbf/in^2

kilogram-force per square metre (kgf/m^2)
m-kgf-s unit of pressure
9.806 65 Pa
1.0 $\times$ **10^{-4}** kgf/cm^2 (q.v)

kilogram-force second (kgf·s)
m-kgf-s unit of momentum
9.806 65 kg·m/s

kilogram-force second per square metre (kgf·s/m^2)
m-kgf-s unit of (dynamic) viscosity
9.806 65 Pa·s

kilogram-force second squared per metre ($kgf{\cdot}s^2/m$)
m-kgf-s unit of mass
9.806 65 kg
Note: cf. metric technical unit of mass

kilogram-force second squared per metre to the fourth power ($kgf{\cdot}s^2/m^4$)
m-kgf-s unit of density
9.806 65 kg/m^3

kilogram-weight (kg*, kg(wt))
obsol. name and symbol for kilogram-force (q.v.)

kilohyl (khyl)
m-kgf-s unit of mass
= metric technical unit of mass (q.v.)

kilohyl per cubic metre ($khyl/m^3$)
unit of density
Note: name sometimes used for kilogram-force second squared per metre to the fourth power (q.v.)

kilometre km
mSI unit of length
1.0 $\times$ $\mathbf{10^3}$ m
$6.213\ 71 \times 10^{-1}$ mile

kilometre per hour km/h
oSI unit of velocity
$2.777\ 78 \times 10^{-1}$ m/s
$6.213\ 71 \times 10^{-1}$ mile/h

kilomole kmol
$\mathbf{10^3}$ mole (q.v.)

kilopond (kp)
m-kp-s unit of force
= kilogram-force (q.v.)
Note: name used in some countries instead of kilogram-force. In all entries in this dictionary ‘kilopond’ may be substituted for ‘kilogram-force’ to change m-kgf-s units to m-kp-s units. Kilopond, now obsolete, was widely used in Central Europe

kilowatt kW
mSI unit of power
10^3 W

kilowatt hour kW·h
kilowattheure
Kilowattstunde
oSI unit of energy
3.6 $\times$ **10^6** J
$8.598\,45\times10^2$ $kcal_{IT}$
$3.412\,14\times10^3$ Btu
$3.670\,98\times10^5$ kgf·m

kip (—)
US unit of force
10^3 lbf$=4.448\,22\times10^3$ N

km kilometre

kmc; KMC kilomegacycles (per second)
=gigahertz (GHz)

kn knot (international)

knot (international) (kn)
nœud
Knoten
unit of velocity
=n mile/h
$5.144\,44\times10^{-1}$ m/s
1.852 km/h
1.150 78 mile/h
$9.993\,61\times10^{-1}$ UK knot

knot (UK) (—)
unit of velocity
$5.147\,73\times10^{-1}$ m/s
1.853 18 km/h
1.000 64 kn (internat. knot)
1.0 UK nautical mile per hour

kp kilopond

kph kilometre per hour (km/h)

kW kilowatt

kWh kilowatt hour (kW·h)

L

l litre

L litre; lambert

l_n *Normliter* (q.v.)

La; la lambert

lambda (λ)
obsol. unit of volume ('capacity')
1 μl or μL

lambert (La)
unit of luminance
3.183 10×10^3 cd/m^2

langley (—)
unit of surface density of radiant energy (radiant exposure)
4.186 8 $\times$ **10^4** J/m^2 = **1** cal$_{IT}$/cm^2
Note: used for measuring radiant energy received from the Sun; also values 1 cal$_{15}$/cm^2 = 41 855 J/m^2 and 1 cal$_{th}$/cm^2 = 41 840 J/m^2 were used

langley per minute langley/min
unit of irradiance
6.978 $\times$ **10^2** W/m^2 or J/(m^2·s) = **1** cal$_{IT}$/(cm^2·s)
Note: used as a unit of insolation (=*in*coming *sol*ar radi*ation*)

l·atm litre atmosphere

lb pound; also $\log_2$

LB pound-force (lbf)

lbf pound-force

lb tr troy pound

li link (—)

light year (l.y.)
année de lumière (a.l.)
Lichtjahr (Lj)
unit of length
$9.460\,7 \times 10^{15}$ m
Note: mostly $9.460\,5 \times 10^{12}$ km is used

link (—)
unit of length
2.011 68 × **10**$^{-1}$ m
7.92 in
1.0 × **10**$^{-2}$ chain
Note: also called Gunter's or surveyor's link

liq oz liquid ounce

liq pt liquid pint

liq qt liquid quart

liquid . . . see also: fluid . . .

liquid ounce (US) (US liq oz)
US unit of volume (capacity) for liquid measure
$2.957\,35 \times 10^{-5}$ m^3
$2.957\,35 \times 10^{-2}$ dm^3 or litre
1.804 69 in^3
1.040 84 UK fl oz
M/S: Table 5
Note: also called fluid ounce

liquid pint (US) (US liq pt)
US unit of volume (capacity) for liquid measure
$4.731\,76 \times 10^{-4}$ m³
$4.731\,76 \times 10^{-1}$ dm³ or litre
$1.671\,01 \times 10^{-2}$ ft³
$2.887\,5 \times 10$ in³
$8.326\,74 \times 10^{-1}$ UKpt
M/S: Table 5

liquid quart (US) (US liq qt)
US unit of volume (capacity) for liquid measure
$9.463\,53 \times 10^{-4}$ m³
M/S: Table 5

litre l, L
litre
Liter
oSI unit of volume
10^{-3} m³ = **1** dm³
Note: the litre was defined by 12th CGPM, 1964, as 1 dm³ exactly; the two symbols l and L are on equal footing – one of them will be suppressed; L is preferred in the USA and Canada; US spelling of this unit is liter; cf. cubic decimetre

litre(old) l
depr. unit of volume (capacity)
$1.000\,028 \times 10^{-3}$ m³
Note: used 1901–1964

litre atmosphere (l·atm)
unit of work (energy)
$1.013\,25 \times 10^{2}$ J

litre per 100 kilometres l/100 km, L/100 km
oSI unit of fuel consumption
$3.540\,06 \times 10^{-3}$ UKgal/mile
$\frac{282.481}{\text{litres}}$ mile/UKgal
$4.251\,44 \times 10^{-3}$ USgal/mile
$\frac{235.215}{\text{litres}}$ mile/USgal

litre per kilogram l/kg, L/kg
oSI unit of specific volume
=cubic decimetre per kilogram (q.v.)

litre per mole l/mol, L/mol
oSI unit of molar volume
$\mathbf{10^{-3}}$ m^3/mol

litre per second l/s, L/s
oSI unit of volume flow rate
$\mathbf{10^{-3}}$ m^3/s

Lj G abbr. for *Lichtjahr*=light year

lm lumen

lumen lm
lumen
Lumen
SI unit of luminous flux
=cd sr

lumen hour lm·h
oSI unit of quantity of light
$\mathbf{3.6 \times 10^3}$ lm·s

lumen per square foot lm/ft^2
unit of illuminance
$1.076\,39 \times 10$ lx

lumen per square metre lm/m^2
SI unit of luminous exitance
=m^{-2} cd sr
=lux (q.v.)

lumen per watt lm/W
SI unit of luminous efficacy
=m^{-2} kg^{-1} s^3 cd sr

lumen second lm·s
SI unit of quantity of light
=s cd sr

lusec (—)
unit of leak rate used in vacuum technology
$1.333\,22 \times 10^{-4}$ N·m/s or W
Note: short for litre (l) at a pressure of 1 micron (μ=u) of mercury per second (sec)

lux lx
lux
Lux
SI unit of illuminance
$= m^{-2}$ cd sr
= lumen per square metre (q.v.)

lux hour lx·h
oSI unit of light exposure
$\mathbf{3.6 \times 10^3}$ lx·s

lux second lx·s
SI unit of light exposure
$= m^{-2}$ s cd sr

lx lux

ly langley (—)

l.y. light year

M

m metre; milli

M mega; maxwell; megabyte (Mbyte)

ma myria

Mach number Ma; M
Mach number is the ratio (v/c) of the velocity (v) of an object or fluid to the velocity of sound (c) in the same medium and under the same conditions. Used e.g. to express the velocity of aircraft. M1 = the velocity of sound, M2 = twice the velocity of sound etc. Velocity of sound in dry air at 0 °C is about 331.46 m/s = 1193.3 km/h

magnetic ohm
Note: name sometimes used for gilbert per maxwell (q.v.)

maxwell Mx; M
CGSm unit of magnetic flux
10^{-8} Wb

maxwell per square centimetre (Mx/cm^2)
=gauss (q.v.)

mb millibar

mbar millibar

mc; Mc megacycle (per second)=megahertz (MHz)

MCM thousand circular mils (cf. circular mil)

ME G symbol for *Masseneinheit* (unit of mass) used for:
(a) *technische Masseneinheit* (see: metric technical unit of mass)
(b) *Masseneinheit*=atomic mass unit (old chemical), q.v.

mechanical ohm (—)
depr. unit of mechanical impedence
10^{-3} N·s/m
Note: name sometimes given to dyne second per centimetre (q.v.)

mega M
SI prefix denoting $\times 10^6$. Examples: megaampere (MA), megabecquerel (MBq), megacalorie (Mcal), megacoulomb (MC), megaelectronvolt (MeV), megaerg (Merg), megagram (Mg), megagray (MGy), megahertz (MHz), megajoule (MJ), meganewton (MN), megapascal (MPa), megapond (Mp), megasiemens (MS), megavolt (MV), megawatt (MW), megohm (MΩ).
Note: when referring to memory capacity the prefix (or its symbol) denotes *approximately* $\times 10^6$, usually $\times 1\,048\,576$ ($=2^{20}$), e.g. megabyte (Mbyte, often Mb)$=2^{20}$ byte

megagram Mg

mSI unit of mass

10^3 kg = **1** t

Note: rarely used; tonne is used in practice

megapond (Mp)

10^3 kilopond (q.v.)

metre m

mètre

Meter

base SI unit of length

Def: see Appendix 1

5.399 57 $\times 10^{-4}$ nautical mile (international)
5.396 12 $\times 10^{-4}$ nautical mile (UK)
6.213 71 $\times 10^{-4}$ mile
1.093 61 yd
3.280 84 ft
3.937 01 $\times 10$ in

Note: US spelling of this unit is meter. Also unit of breadth, height, depth, thickness, radius, diameter, length of path, wavelength, sound particle displacement, mean free path, stopping equivalent, mean linear range, diffusion coefficient for neutron fluence rate, Burgers vector, displacement vector of ion, equilibrium position vector of ion, fundamental lattice vector, half-thickness, lattice vector and particle position vector

metre cubed m^3

mètre cube

Kubikmeter; (*Meter hoch drei*)

SI unit of section modulus

Note: the same as cubic metre (q.v.)

metre hour degree Celsius per kilocalorie (I.T.) ($\mathrm{m \cdot h \cdot °C/kcal_{IT}}$)

unit of thermal resistivity

8.598 45 $\times 10^{-1}$ m·K/W
3.6 $\times$ **10**^{2} $\mathrm{cm \cdot s \cdot °C/cal_{IT}}$
1.488 16 ft·h·°F/Btu

metre kelvin m·K
SI unit of second radiation constant

metre kelvin per watt m·K/W
SI unit of thermal resistivity
$=m^{-1}\ kg^{-1}\ s^{3}\ K = m{\cdot}s{\cdot}K/J$
4.186 8 ×**10**^{2} cm·s·°C/cal_{IT}
1.163 m·h·°C/$kcal_{IT}$
1.730 73 ft·h·°F/Btu

metre of water (conventional) (mH_2O)
unit of pressure
9.806 65×**10**^{3} Pa

metre per kilogram m/kg
SI unit of specific length
1.488 16 ft/lb

metre per second m/s
SI unit of velocity (speed)
3.6 km/h
6.0 ×**10** m/min
3.280 84 ft/s
2.236 94 mile/h
1.943 84 kn (international knot)
1.942 60 UK knot

metre per second cubed m/s^3
SI unit of jerk
3.280 84 ft/s^3

metre per second squared m/s^2
SI unit of acceleration
10^2 Gal = 3.280 84 ft/s^2

metre second per kilogram m·s/kg
= reciprocal pascal reciprocal second (q.v.)

metre squared m^2

SI unit of slowing-down area, diffusion area and migration area

Note: the same as square metre (q.v.)

metre squared per hour m^2/h

oSI unit of kinematic viscosity

$2.777\,78 \times 10^{-4}$ m^2/s

$2.777\,78 \times 10^{2}$ cSt

metre squared per newton second $m^2/(N \cdot s)$

= reciprocal pascal reciprocal second (q.v.)

metre squared per second m^2/s

SI unit of kinematic viscosity

1.0 $\times$ **10^4** St (stokes) or cm^2/s

1.0 $\times$ **10^6** cSt

3.6 $\times$ **10^3** m^2/h

$1.076\,39 \times 10$ ft^2/s

$3.875\,01 \times 10^{4}$ ft^2/h

$1.550\,00 \times 10^{3}$ in^2/s

$5.580\,01 \times 10^{6}$ in^2/h

Note: also unit of thermal diffusivity, diffusion coefficient and thermal diffusion coefficient

metre to the fourth power m^4

SI unit of second moment of area

10^8 cm^4 = **10^{12}** mm^4

$1.158\,62 \times 10^{2}$ ft^4

$2.402\,51 \times 10^{6}$ in^4

metric carat (—)

carat métrique

metrisches Karat

unit of mass

2×10^{-4} kg = **2×10^{-1}** g = **2×10^{2}** mg

Note: adopted by 4th CGPM, 1907, for diamonds, fine pearls and precious stones; not to be confused with carat (karat) used as a measure of purity of gold

metric technical unit of mass (—)
unité technique métrique de masse
technische Masseneinheit (TME or ME)
m-kgf-s unit of mass
1 kgf·s²/m or kp·s²/m
9.806 65 kg
2.162 00 lb
6.719 69 × 10⁻¹ slug
Note: sometimes called kilohyl

mev megaelectronvolt (MeV)

mf millifarad (mF); microfarad (μF)

mF millifarad

mfd microfarad (μF)

. . .ᵐᵍ milligrade

mg milligram

mGal; mgal milligal (mGal)

mho name sometimes used in practice for reciprocal ohm (q.v.)

mi US abbr. for mile. Consequently mi² = square mile and mi³ = cubic mile

micro μ
SI prefix denoting × 10⁻⁶. Examples: microampere (μA), microbar (μbar), microcoulomb (μC), microcurie (μCi), microfarad (μF), microgram (μg), microgray (μGy), microhenry (μH), microinch (μin), microjoule (μJ), microlumen (μlm), micrometre (μm), micromole (μmol), micronewton (μN), microohm (μΩ), micropascal (μPa), microradian (μrad), microsecond (μs), microsiemens (μS), microtesla (μT), microvolt (μV), microwatt (μW)

microbar μbar
$= \mathbf{10^{-1}}$ Pa

micro-inch μin
unit of length
$\mathbf{2.54 \times 10^{-8}}$ m
$\mathbf{1.0 \times 10^{-6}}$ in

microkatal μkat
unit of enzyme activity
1 μmol/s
Note: 1 μmol of reaction product produced or consumed per second

micrometre μm
micromètre
Mikrometer
mSI unit of length
$\mathbf{10^{-6}}$ m
Note: formerly called micron

micron (μ)
micron
Mikron
Note: obsol., now called micrometre (q.v.)

micron of mercury μHg
obsol. unit of fluid pressure
$1.333\,22 \times 10^{-1}$ Pa

microtorr μTorr
obsol. unit of fluid pressure
$1.333\,22 \times 10^{-4}$ Pa

mil term formerly used in English speaking countries for various units, including: milli-inch, millilitre and milliradian, and also $360°/6400 = 0.05625° = 9.817\,48 \times 10^{-4}$ rad
Note: cf. circular mil

mile mile

UK and US unit of length

1.609 344 $\times$ **10**³ m

1.76 $\times$ **10**³ yd

5.28 $\times$ **10**³ ft

M/S: Table 1

Note: also known as statute mile. Abbr. mi is used in the US. See also: nautical mile (international), nautical mile (UK) and telegraph nautical mile

mile per gallon (UK) mile/UKgal

unit of reciprocal fuel consumption

$3.540\,06 \times 10^{-1}$ km/litre

$\frac{282.481}{\text{miles}}$ litres/100 km

$8.326\,74 \times 10^{-1}$ mile/USgal

mile per gallon (US) mile/USgal

unit of reciprocal fuel consumption

$4.251\,44 \times 10^{-1}$ km/litre

$\frac{235.215}{\text{miles}}$ litres/100 km

1.200 95 mile/UKgal

mile per hour mile/h

UK and US unit of velocity

4.470 4 $\times$ **10**$^{-1}$ m/s

1.609 344 km/h

1.466 67 ft/s

milli m

SI prefix denoting $\times 10^{-3}$. Examples: milliampere (mA), millibar (mbar), millicandela (mcd), millicoulomb (mC), millicurie (mCi), millidarcy (mD), millifarad (mF), milligal (mgal), milligrade (. . . .^{mg}), milligram (mg), milligray (mGy), millihenry (mH), millijoule (mJ), millikelvin (mK), millilitre (ml, mL), millilumen (mlm), millimetre (mm), millimole (mmol), millinewton (mN), milliohm (mΩ), millipascal (mPa), millipièze (mpz), millipond (mp), milliradian (mrad), millisecond (ms), millisiemens (mS), millitesla (mT), millivolt (mV), milliwatt (mW), milliweber (mWb)

millibar mbar; mb
unit of pressure
10^{-3} bar = **10^{2}** Pa
Note: used in meteorological barometry

milligal mGal
unit of acceleration
10^{-5} m/s² = **10^{-3}** Gal

milligrade ....^{mg}
milligrade
Milligon
unit of plane angle
10^{-3} grade (q.v.)

milligram per litre mg/l, mg/L
oSI unit of (mass) density and concentration
10^{-3} kg/m³

milli-inch (min)
unit of length
2.54 × **10^{-5}** m
1.0 × **10^{-3}** in
Note: sometimes called mil or thou

millilitre ml, mL
millilitre
Milliliter
oSI unit of volume (capacity)
10^{-6} m³
1 cm³
10^{-3} litre

millimetre mm
millimètre
Millimeter
mSI unit of length
10^{-3} m
Note: used as fundamental unit in engineering

millimetre of mercury (conventional) (mmHg)
millimètre de mercure (conventionnel)
Millimeter Quecksilbersäule (konventionelle)
unit of pressure
$1.333\,22 \times 10^{2}$ Pa
$1.333\,22 \times 10^{-3}$ bar
1.000 00 torr

millimetre of water (conventional) (mmH_2O)
millimètre d'eau (conventionnel)
Millimeter Wassersäule (konventionelle)
unit of pressure
9.806 65 Pa
$9.806\,65 \times 10^{-5}$ bar
1.0 kgf/m^2 or kp/m^2

millimicron mμ
obsol., now called nanometre (nm)

millitorr mTorr
obsol. unit of fluid pressure
$1.333\,22 \times 10^{-1}$ Pa

min minute; milli-inch; minim (UK or US)

minim (UK) (UKmin)
UK unit of volume (capacity)
$5.919\,39 \times 10^{-8}$ m^3
$5.919\,39 \times 10$ mm^3
$3.612\,23 \times 10^{-3}$ in^3
M/S: Table 4

minim (US) (USmin)
US unit of volume (capacity) for liquid measure
$6.161\,15 \times 10^{-8}$ m^3
$6.161\,15 \times 10$ mm^3
$3.759\,77 \times 10^{-3}$ in^3
M/S: Table 5

minute min
minute
Minute
oSI unit of time
6.0 $\times$**10** s
$1.666\,67\times10^{-2}$ h
$6.944\,44\times10^{-4}$ d
Note: symbol (mn) formerly used in France

minute . . . .
minute
Minute; Altminute
oSI unit of plane angle
$2.908\,88\times10^{-4}$ rad (=π/10 800)
$1.666\,67\times10^{-2}$ ° (degree) (=1/60)
6.0 **10**^{2} ″ (second)
$1.851\,85\times10^{-2}$ ^{g} (grade) (=400/21 600)
$1.851\,85\times10^{-4}$ ∟ (right angle)

minute see: centesimal minute

ml, mL millilitre

mm millimetre

mmHg conventional millimetre of mercury

mmH$_2$O conventional millimetre of water

mmQS G abbr. for *Millimeter Quecksilbersäule* = (conventional) millimetre of mercury (mmHg)

mmWS G abbr. for *Millimeter Wassersäule* = (conventional) millimetre of water (mmH$_2$O)

mol mole

mole mol
mole
Mol
base SI unit of amount of substance
Def: see Appendix 1

mole per cubic metre mol/m^3
SI unit of concentration

mole per kilogram mol/kg
SI unit of molality and ionic strength

mole per litre mol/l, mol/L
oSI unit of concentration
$10^3\ mol/m^3 = 1\ mol/dm^3$

mole per second mol/s
SI unit of molar flow rate

m.p.g. miles per gallon (mile/gal)

m.p.h. miles per hour (mile/h)

ms millisecond

mu micro (μ)

mu, Mu micron = micrometre (μm)

Mx maxwell

myria (ma)
depr. prefix denoting $\times 10^4$

N

n nano

N newton; neper (Np)

nano n
SI prefix denoting $\times 10^{-9}$. Examples: nanoampere (nA), nanocoulomb (nC), nanofarad (nF), nanohenry (nH), nanometre (nm), nanoohm (nΩ), nanosecond (ns), nanosiemens (nS), nanotesla (nT), nanovolt (nV), nanowatt (nW)

nanogram per pascal second square metre ng/(Pa·s·m^2)
mSI unit used in calculations of moisture transport in buildings

nanometre nm
mSI unit of length
10^{-9} m
Note: used to be called millimicron

nautical mile (international) (n mile)
mille marin; mille (—)
Seemeile (sm)
unit of length
1.852 ×10^3 m
1.852 km
9.993 61 × 10^{-1} nautical mile (UK)

nautical mile (UK) (—)
UK unit of length
1.853 184 ×10^3 m
1.853 184 km
1.000 64 n mile (international)
6.08 ×10^3 ft

neper Np
néper
Neper
unit of dimensionless quantities: amplitude level difference, power level difference and logarithmic decrement
$n = k \ln (Q_1/Q_2)$
where n = number of nepers, Q_1 and Q_2 = quantities of the same kind, k = 0.5 or 1 depending on the quantities. E.g. for power level difference k = 0.5, for amplitude level difference k = 1
1 Np = 8.685 89 dB (= 20 lg e)

neper per second Np/s
unit of damping coefficient
1 s^{-1}

newton N
newton
Newton
SI unit of force and 'weight'
$= \text{m kg s}^{-2}$
1.0 $\times$ **10^5** dyn
1.0 $\times$ **10^{-3}** sn
$1.019\,72 \times 10^{-1}$ kgf or kp
7.233 01 pdl
$2.248\,09 \times 10^{-1}$ lbf

newton metre N·m
mètre-newton
Newtonmeter
SI unit of moment of force and torque
$= \text{m}^2 \text{ kg s}^{-2} = \text{J}$
1.0 $\times$ **10^7** dyn·cm
$1.019\,72 \times 10^{-1}$ kgf·m or kp·m
$2.373\,04 \times 10$ pdl·ft
$7.375\,62 \times 10^{-1}$ lbf·ft
8.850 75 lbf·in

newton metre per second N·m/s
=pascal cubic metre per second (q.v.)

newton metre second N·m·s
=kilogram metre squared per second (q.v.)

newton metre squared per ampere $\text{N·m}^2/\text{A}$
=weber metre (q.v.)

newton per cubic metre N/m^3
unit of 'specific weight'
$= \text{m}^{-2} \text{ kg s}^{-2}$
10^{-1} dyn/cm^3

newton per metre N/m
SI unit of surface tension
$= \text{kg s}^{-2} = \text{J/m}^2$
1.0 $\times$ **10^3** dyn/cm or erg/cm^2
$1.019\,72 \times 10^{-1}$ kgf/m or kp/m
Note: also unit of force per unit length

newton per metre cubed N/m^3
=pascal per metre (q.v.)

newton per square metre N/m^2
=pascal (q.v.)

newton per weber N/Wb
=ampere per metre (q.v.)

newton second N·s
=kilogram metre per second (q.v.)

newton second per metre N·s/m
SI unit of mechanical impedance
=kg s^{-1}
1.0 ×**10**^{3} dyn·s/m

newton second per metre cubed N·s/m^3
=pascal second per metre (q.v.)

newton second per metre squared N·s/m^2
=pascal second (q.v.)
Note: formerly called *poiseuille* (Pl) in France

newton second per metre to the fifth power N·s/m^5
=pascal second per metre cubed (q.v.)

newton second per square metre N·s/m^2
=newton second per metre squared (q.v.)

newton square metre per ampere N·m^2/A
=weber metre (q.v.)

newton square metre per kilogram squared N·m^2/kg^2
SI unit of gravitational constant
=m^3 kg^{-1} s^{-2}

nile (—)
unit of dimensionless quantity: reactivity
=**10**$^{-2}$
Note: its sub-multiple millinile=**10**$^{-5}$ is usually used

nit (nt)
unit of luminance
=candela per square metre (q.v.)

Nm³ *Normkubikmeter* (q.v.)

n mile international nautical mile

normal atmosphere see: standard atmosphere (atm)

Normkubikmeter Nm^3
G obsol. unit
1 m^3 of gas under standard reference conditions
Note: formerly used also in other Continental countries

Normliter l_n
G obsol. unit
1 litre of gas under standard reference conditions

nox (nx)
obsol. unit of scotopic illuminance
10^{-3} lx

Np neper

nt nit

nx nox

O

octant (—)
unit of plane angle
$(\pi/4)$ rad = 45°

octave (—)
unit of dimensionless quantity: frequency interval
Def: The interval between two frequencies f_1, f_2 having a ratio $f_1/f_2 = 2$
$= \mathrm{lb}\,(f_1/f_2)$
301 savart = 1200 cent

octet see: byte

oersted (Oe)
CGSm unit of magnetic field strength
=Gb/cm
$7.957\,75 \times 10$ A/m [$=10^3/(4\pi)$]

ohm Ω
SI unit of resistance, impedance, modulus of impedance and reactance
$= m^2\ kg\ s^{-3}\ A^{-2} = V/A = W/A^2 = S^{-1}$

ohm circular mil per foot (Ω·circ·mil/ft)
unit of resistivity
$1.662\,43 \times 10^{-9}$ Ω·m

ohm metre Ω·m
SI unit of resistivity
$m^3\ kg\ s^{-3}\ A^{-2}$
10^6 $\Omega\cdot mm^2/m$

ohm second Ω·s
=henry (q.v.)

ohm square millimetre per metre $\Omega\cdot mm^2/m$
unit of resistivity
10^{-6} Ω·m

ounce (oz)
UK and US unit of mass
$2.834\,95 \times 10^{-2}$ kg
$2.834\,95 \times 10$ g
4.375 $\times$ **10^2** gr
M/S: Table 7
Note: this is an avoirdupois unit

ounce, apothecaries' (oz apoth; oz ap)
UK and US unit of mass
=**1** ounce, troy (q.v.)
Note: the abbreviation oz apoth is used in UK, and oz ap in US; see 'apothecaries' units' and Table 10

ounce, avoirdupois see: ounce

ounce, fluid see: fluid ounce

ounce, liquid see: liquid ounce

ounce, troy (oz tr)
UK and US unit of mass
$3.110\,35 \times 10^{-2}$ kg
1.0 oz apoth
4.8 $\times$ **10**^{2} gr
Note: see 'troy units' and Table 9; not lawful for trade in UK except for the purposes of transactions in, or in articles made from, gold, silver or other precious metals, including transactions in gold or silver thread, lace or fringe

ounce inch squared oz·in^2
unit of moment of inertia
$1.829\,00 \times 10^{-5}$ kg·m^2
$4.340\,28 \times 10^{-4}$ lb·ft^2

ounce per cubic inch oz/in^3
unit of (mass) density
$1.729\,99 \times 10^3$ kg/m^3

ounce per foot oz/ft
unit of linear density
$9.301\,02 \times 10^4$ kg/m

ounce per gallon (UK) oz/UKgal
UK unit of (mass) density and concentration
6.236 02 kg/m^3

ounce per gallon (US) oz/USgal
US unit of (mass) density and concentration
7.489 15 kg/m^3

ounce per inch oz/in
unit of linear density
1.116 12 kg/m

ounce per square foot oz/ft²
unit of surface density
$3.051\,52 \times 10^{-1}$ kg/m²
$3.051\,52 \times 10^{2}$ g/m²
9.0 oz/yd²
6.25 $\times$ **10** lb/1000 ft²
2.7225 $\times$ **10**³ lb/acre

ounce per square yard oz/yd²
unit of surface density
$3.390\,57 \times 10^{-2}$ kg/m²
$3.390\,57 \times 10$ g/m²
$1.111\,11 \times 10^{-1}$ oz/ft²
6.944 44 lb/1000 ft²
3.025 $\times$ **10**² lb/acre

ounce per yard oz/yd
unit of linear density
$3.100\,34 \times 10^{-2}$ kg/m

ounce-force (ozf)
unit of force
$2.780\,14 \times 10^{-1}$ N
$2.834\,95 \times 10^{-2}$ kgf or kp
2.010 88 pdl
6.25 $\times$ **10**⁻² lbf

ounce-force inch (ozf in)
unit of moment of force and torque
$7.061\,55 \times 10^{-3}$ N·m
$7.200\,78 \times 10^{-4}$ kgf·m or kp·m
$5.208\,33 \times 10^{-3}$ lbf·ft

ounce-force per square inch (ozf/in²)
unit of pressure
$4.309\,22 \times 10^{2}$ Pa

oz ounce (avoirdupois)

oz ap; oz apoth apothecaries' ounce

ozf ounce-force

oz t; oz tr troy ounce

P

p pico; pond

P peta; poise

Pa pascal

parsec pc
parsec
Parsec
oSI unit of length
$3.085\,7 \times 10^{16}$ m
$2.062\,66 \times 10^{5}$ astronomic unit
Note: formed by blending from '*par*allax of one *sec*ond'; also multiples formed by SI prefixes are used

pascal Pa
pascal
Pascal
SI unit of pressure and stress
$= m^{-1}\ kg\ s^{-2} = N/m^2 = J/m^3$
$= \mathbf{10}^{-5}$ bar $= \mathbf{10}$ μbar or dyn/cm²
$1.019\,72 \times 10^{-1}$ kgf/m² or kp/m²
$1.019\,72 \times 10^{-5}$ kgf/cm² or kp/cm² or at
$9.869\,23 \times 10^{-6}$ atm
$7.500\,62 \times 10^{-3}$ Torr
$6.719\,69 \times 10^{-1}$ pdl/ft²
$2.088\,54 \times 10^{-2}$ lbf/ft²
$1.450\,38 \times 10^{-4}$ lbf/in²
Note: also unit of bulk modulus, fugacity, modulus of elasticity, and shear modulus

pascal cubic metre Pa·m³
SI unit of quantity of gas
$= m^2\ kg\ s^{-2} = J = N\cdot m$

pascal cubic metre per second Pa·m³/s
SI unit of throughput and leak rate
$=m^2 kg s^{-3} = N·m/s = J/s = W$

pascal litre Pa·l, Pa·L
oSI unit of quantity of gas
10^3 Pa·m³

pascal litre per second Pa·l/s, Pa·L/s
oSI unit of throughput
10^3 Pa·m³/s

pascal per kelvin Pa/K
SI unit of pressure coefficient
$=m^{-1} kg s^{-2} K^{-1}$

pascal per metre Pa/m
SI unit of pressure gradient
$=m^{-2} kg s^{-2}$

pascal second Pa·s
SI unit of (dynamic) viscosity
$=m^{-1} kg s^{-1} = N·s/m^2$
10 P (poise)

pascal second per metre Pa·s/m
SI unit of characteristic impedance of a medium and specific acoustic impedance
$=m^{-2} kg s^{-1} = N·s/m^3$
10^{-1} dyn·s/cm³

pascal second per metre cubed Pa·s/m³
SI unit of acoustic impedance
$=m^{-4} kg s^{-1} = N·s/m^5$
10^{-5} dyn·s/cm⁵

pc parsec; per cent (%)

pcm pour cent mille

pdl poundal

peck (—)
UK unit of volume (capacity)
$9.092\,18 \times 10^{-3}$ m^3
M/S:Table 4

peck (pk)
US unit of volume (capacity) for dry measure
$8.809\,77 \times 10^{-3}$ m^3
M/S: Table 6

pennyweight (dwt)
unit of mass
$1.555\,17 \times 10^{-3}$ kg
Note: see 'troy units' and Table 9

per ... (e.g. per second) – see: reciprocal ...

per cent %
pour cent
Perzent
$1/100 = 10^{-2}$
Note: see Part 2: mass per cent, mole per cent, volume per cent

per thousand ‰
pour mille
Promille
$1/1000 = 10^{-3}$

perch (—)
depr. unit of length
=rod (q.v.)

perch (p)
depr. unit of area
$2.529\,29 \times 10$ m^2 (=square rod)

peta P
SI prefix denoting $\times 10^{15}$. Examples: petabecquerel (PBq), petahertz (PHz), petajoule (PJ), petametre (Pm), petaohm (PΩ), petawatt (PW)

Pfd Pfund (q.v.)

Pfund Pfd
G obsol. unit of mass
0.5 kg

ph phot

phon (—)
phone
Phon
unit of dimensionless quantity: loudness level
Note: see sone

phot (ph)
unit of illuminance
$\mathbf{10^4}$ lx

phot-second (ph·s)
unit of light exposure
$\mathbf{10^4}$ lx·s

pico p
SI prefix denoting $\times 10^{-12}$. Examples: picoampere (pA), picocoulomb (pC), picofarad (pF), picohenry (pH), picometre (pm), picosecond (ps), picowatt (pW)

pièze (pz)
MTS unit of pressure
$\mathbf{10^3}$ Pa

pint (UK) (UKpt)
UK unit of volume (capacity)
$5.682\,61 \times 10^{-4}$ m^3
$5.682\,61 \times 10^{-1}$ dm^3 or litre
$2.006\,79 \times 10^{-2}$ ft^3
$3.467\,74 \times 10$ in^3
1.200 95 US liq pt
M/S: Table 4

pint see: dry pint, liquid pint

pk US peck

Pk G abbr. for *Parsec* = parsec

Pl poiseuille

point
(a) unit of mass
10^{-2} metric carat = **2** mg
(b) obsol. unit of (plane) angle
360°/32 = 11.25°

poise P
poise
Poise
CGS unit of (dynamic) viscosity
= dyn·s/cm^2 = g/(cm·s)
10^{-1} pascal second (q.v.)

poiseuille Pl
unit of (dynamic) viscosity
= pascal second (q.v.)
Note: formerly used in France; cf. poise

pole (—)
depr. unit of length
= rod (q.v.)

poncelet (—)
F obsol. unit of power
9.806 65 × **10**^{2} W

pond (p)
unit of force
10^{-3} kilopond (q.v.)

pound lb

UK and US unit, and FPS base unit of mass
Def: the pound equals to 0.453 592 37 kilogram exactly (from 1959 in US, from 1963 in UK)
$4.535\,92 \times 10^{-1}$ kg
$4.625\,35 \times 10^{-2}$ metric technical unit of mass
$3.108\,10 \times 10^{-2}$ slug
$4.464\,29 \times 10^{-4}$ ton (UK or long)
$\mathbf{5.0 \times 10^{-4}}$ short ton
$\mathbf{7.0 \times 10^{3}}$ gr
M/S: Table 7
Note: this is an avoirdupois unit

pound, troy (lb tr)

depr. unit of mass
$3.732\,42 \times 10^{-1}$ kg
Note: see 'troy units' and Table 9

pound foot per second lb·ft/s

FPS unit of momentum
$1.382\,55 \times 10^{-1}$ kg·m/s

pound foot per second squared lb·ft/s^2

=poundal (q.v.)

pound foot squared lb·ft^2

FPS unit of moment of inertia
$4.214\,01 \times 10^{-2}$ kg.m^2

pound foot squared per second lb·ft^2/s

FPS unit of moment of momentum
$4.214\,01 \times 10^{-2}$ kg·m^2/s

pound inch squared lb·in^2

unit of moment of inertia
$2.926\,40 \times 10^{-4}$ kg·m^2
$6.944\,44 \times 10^{-3}$ lb·ft^2

pound per acre lb/acre
unit of surface density
1.120 85 $\times 10^{-4}$ kg/m^2
1.120 85 kg/ha
2.295 68 $\times 10^{-2}$ lb/1000 ft^2
3.305 79 $\times 10^{-3}$ oz/yd^2
3.673 09 $\times 10^{-4}$ oz/ft^2

pound per cubic foot lb/ft^3
FPS unit of (mass) density
1.601 85 $\times 10$ kg/m^3
5.787 04 $\times 10^{-4}$ lb/in^3
1.605 44 $\times 10^{-1}$ lb/UKgal
1.336 81 $\times 10^{-1}$ lb/USgal

pound per cubic inch lb/in^3
unit of (mass) density
2.767 99 $\times 10^{4}$ kg/m^3
1.728 $\times$ **10**^{3} lb/ft^3

pound per foot lb/ft
FPS unit of linear density
1.488 16 kg/m
8.333 33 $\times 10^{-2}$ lb/in
3.0 lb/yd

pound per foot second lb/(ft·s)
=poundal second per square foot (q.v.)

pound per gallon (UK) (lb/UKgal)
UK unit of (mass) density
9.977 64 $\times 10$ kg/m^3
6.228 84 lb/ft^3
8.326 74 $\times 10^{-1}$ lb/USgal

pound per gallon (US) (lb/USgal)
US unit of (mass) density
1.198 26 $\times 10^{2}$ kg/m^3
7.480 52 lb/ft^3
1.200 95 lb/UKgal

pound per hour lb/h
unit of mass flow rate
$1.259\,98 \times 10^{-4}$ kg/s
$4.535\,92 \times 10^{-1}$ kg/h
$2.777\,78 \times 10^{-4}$ lb/s

pound per inch lb/in
unit of linear density
$1.785\,80 \times 10$ kg/m
1.2 $\times$ **10** lb/ft
3.6 $\times$ **10** lb/yd

pound per second lb/s
FPS unit of mass flow rate
$4.535\,92 \times 10^{-1}$ kg/s
$1.632\,93 \times 10^{3}$ kg/h
3.6 $\times$ **10**^{3} lb/h

pound per square foot lb/ft^2
FPS unit of surface density
4.882 43 kg/m^2
Note: this name and symbol are often used incorrectly for pound-force per square foot (q.v.)

pound per square inch lb/in^2
unit of surface density
$7.030\,70 \times 10^{2}$ kg/m^2
Note: this name and symbol are often used incorrectly for pound-force per square inch (q.v.)

pound per square yard lb/yd^2
unit of surface density
$5.424\,92 \times 10^{-1}$ kg/m^2

pound per thousand square feet lb/1000 ft^2
unit of surface density
$4.882\,43 \times 10^{-3}$ kg/m^2
$4.882\,43 \times 10$ kg/ha
4.356 $\times$ **10** lb/acre
1.44 $\times$ **10**$^{-1}$ oz/yd^2
1.6 $\times$ **10**$^{-2}$ oz/ft^2

pound per yard lb/yd
unit of linear density
4.960 55 × 10^{-1} kg/m
2.777 78 × 10^{-2} lb/in
3.333 33 × 10^{-1} lb/ft

poundal pdl
FPS unit of force
= lb·ft/s^2
1.382 55 × 10^{-1} N
1.409 81 × 10^{-2} kgf or kp
3.108 10 × 10^{-2} lbf

poundal foot pdl·ft
FPS unit of moment of force and torque
4.214 01 × 10^{-2} N·m or J
3.108 01 × 10^{-2} lbf·ft
Note: cf. foot poundal

poundal per square foot pdl/ft^2
FPS unit of pressure
1.488 16 Pa
1.517 50 × 10^{-5} kgf/cm^2 or kp/cm^2
3.108 10 × 10^{-2} lbf/ft^2

poundal second per square foot pdl·s/ft^2
FPS unit of (dynamic) viscosity
1.488 16 Pa·s
1.0 lb/ft·s

pound-force (lbf)
ft-lbf-s base unit of force
4.448 22 N
4.535 92 × 10^{-1} kgf
3.217 40 × 10 pdl

pound-force foot (lbf·ft)
ft-lbf-s unit of moment of force and torque
1.355 82 N·m
1.382 55 × 10^{-1} kgf·m or kp·m

pound-force hour per square foot (lbf·h/ft²)
unit of (dynamic) viscosity
$1.723\,69 \times 10^{5}$ Pa·s

pound-force inch (lbf·in)
unit of moment of force and torque
$1.129\,85 \times 10^{-1}$ N·m
2.681 17 pdl·ft
$8.333\,33 \times 10^{-2}$ lbf·ft

pound-force per foot (lbf/ft)
ft-lbf-s unit of surface tension
$1.459\,39 \times 10$ N/m

pound-force per inch (lbf/in)
unit of surface tension
$1.751\,27 \times 10^{2}$ N/m

pound-force per square foot (lbf/ft²)
ft-lbf-s unit of pressure
$4.788\,03 \times 10$ Pa
$4.882\,43 \times 10^{-4}$ kgf/cm² or kp/cm²
$6.944\,44 \times 10^{-3}$ lbf/in²

pound-force per square inch (lbf/in²)
unit of pressure
$6.894\,76 \times 10^{3}$ Pa
$7.030\,70 \times 10^{-2}$ kgf/cm² or kp/cm²
$\mathbf{1.44 \times 10^{2}}$ lbf/ft²

pound-force second per square foot (lbf·s/ft²)
ft-lbf-s unit of (dynamic) viscosity
$4.788\,03 \times 10$ Pa·s
1.0 slug/(ft·s)

pound-weight (Lb)
Note: other name for pound-force (q.v.)

pour cent mille pcm
unit of dimensionless quantity : reactivity
$= \mathbf{10^{-5}}$ (cf. nile)

ppb parts per billion (US) $= 10^{-9}$

ppM parts per milliard $= 10^{-9}$

ppm parts per million $= 10^{-6}$

PS G abbr. for *Pferdestärke* = metric horsepower (—)

p.s.f.; psf 'pounds' per square foot, correctly: pound-force per square foot (lbf/ft^2)

p.s.i.; psi 'pounds' per square inch, correctly: pound-force per square inch (lbf/in^2)

psia 'pounds' per square inch (absolute) – incorrect; instead of 'pressure of x psia' write 'absolute pressure of x lbf/in^2'

psig 'pounds' per square inch (gauge) – incorrect; instead of 'pressure of x psig' write 'gauge pressure of x lbf/in^2'

pt, pt. pint

pz pièze

Q

q quintal; G depr. abbr. for *Quadrat-* (e.g. qcm = cm^2)

qm G obsol. abbr. for *Quadratmeter* = square metre (m^2)

qt UK quart

quad (—)
US unit of heat energy (of fuel reserves)
1.055×10^{18} J = 1.055 EJ
Note: short for '*quad*rillion Btu' (= 10^{15} Btu)

quart (UK) (UKqt)
UK unit of volume (capacity)
$1.136\,52 \times 10^{-3}$ m³

quart see: dry quart, liquid quart

quarter (qr)
UK unit of mass
$1.270\,06 \times 10$ kg
2.8 × **10** lb
M/S: Table 7
Note: colloquially quarter is short for 'quarter of a pound' = **4** oz = 113.4 g

quintal (q)
quintal (q)
Dezitonne (dt)
unit of mass
10^2 kg

Q-unit (—)
US unit of heat energy (of fuel reserves)
1.055×10^{21} J = 1.055×10^3 EJ
Note: short for 'quintillion Btu' (= 10^{18} Btu)

R

r revolution

R röntgen

°R degree Rankine; (degree Réaumur)

rad rad; radian

rad rad, rd
rad
Rad
unit of absorbed dose
10^{-2} Gy = **10^2** erg/g
Note: the symbol for this unit is the same as the symbol of the unit radian; the symbol rd may be used to avoid confusion

rad per second rad/s, rd/s
unit of absorbed dose rate
10^{-2} Gy/s = **10^{2}** erg/(g·s)

radian rad
radian
Radiant
supplementary SI unit of (plane) angle
Def: the radian is the plane angle between two radii of a circle which cut off on the circumference an arc equal in length to the radius
57° 17′ 44.8″
5.729 58 × 10 ° (degree) (=180/π)
3.437 75 × 10^{3} ′ (minute)
2.062 65 × 10^{5} ″ (second) (=206 264.806″)
6.366 20 × 10 ^{g} (grade) (=200/π)
6.366 20 × 10^{-1} ∟ (right angle)
Note: also SI unit of phase difference

radian per metre rad/m
unit of phase coefficient
= m^{-1}

radian per minute rad/min
oSI unit of angular velocity
1.666 67 × 10^{-2} rad/s
9.549 30 × 10^{-1} °/s

radian per second rad/s
SI unit of angular velocity and circular frequency
5.729 58 × 10 °/s
6.366 20 × 10 ^{g}/s

radian per second squared rad/s^2
SI unit of angular acceleration
5.729 58 × 10 °/s^2
6.366 20 × 10 ^{g}/s^2

rayl (—)
obsol. unit of specific acoustic impedance
10 Pa·s/m

rd F obsol. symbol for radian; symbol for rad

rd; Rd rutherford

reciprocal angström $Å^{-1}$
unit of wavenumber
$\mathbf{10^{10}}$ $m^{-1} = \mathbf{10^{8}}$ cm^{-1}

reciprocal centimetre cm^{-1}
mSI and CGS unit of wavenumber
$\mathbf{10^{2}}$ m^{-1}
Note: used in spectroscopy

reciprocal cubic metre m^{-3}
SI unit of number density (e.g. acceptor n.d., donor n.d., electron n.d., hole n.d., intrinsic n.d., ion n.d., neutron n.d., n.d. of molecules or particles) and molecular concentration

reciprocal cubic metre reciprocal second $m^{-3} \cdot s^{-1}$
SI unit of volume collision rate

reciprocal degree Celsius $°C^{-1}$
same as reciprocal kelvin (q.v.)

reciprocal electronvolt reciprocal cubic metre $eV^{-1} \cdot m^{-3}$
oSI unit of density of states
$6.241\,46 \times 10^{18}$ $J^{-1} \cdot m^{-3}$

reciprocal farad F^{-1}
SI unit of reciprocal capacitance (=elastance)

reciprocal henry H^{-1}
SI unit of reluctance
$= m^{-2}\ kg^{-1}\ s^{2}\ A^{2} = A/Wb = A/(V \cdot s) = S/s$

reciprocal joule reciprocal cubic metre $J^{-1} \cdot m^{-3}$
SI unit of density of states
$= m^{-5}\ kg^{-1}\ s^{2}$

reciprocal kelvin K^{-1}
SI unit of linear expansion coefficient
$= °C^{-1}$
$5.555\,55 \times 10^{-1}\,°F^{-1}$
Note: also unit of cubic expansion coefficient and relative pressure coefficient

reciprocal metre m^{-1}
SI unit of wavenumber and circular wavenumber
$\mathbf{10^{-2}}\,cm^{-1} = \mathbf{1}$ dioptre
Note: also SI unit of attenuation coefficient, linear absorption coefficient, linear attenuation coefficient, linear ionization, macroscopic cross section, phase coefficient, power of a lens, propagation coefficient and Rydberg constant

reciprocal minute min^{-1}
oSI unit of circular frequency
$1.666\,67 \times 10^{-2}\,s^{-1}$

reciprocal mole mol^{-1}
SI unit of Avogadro constant

reciprocal nanometre nm^{-1}
mSI unit of wavenumber
$\mathbf{10^{9}}\,m^{-1}$

reciprocal ohm Ω^{-1}
= siemens (q.v.)

reciprocal ohm metre $1/(\Omega \cdot m)$
= siemens per metre (q.v.)
Note: the correct name is reciprocal ohm reciprocal metre ($\Omega^{-1} \cdot m^{-1}$)

reciprocal pascal Pa^{-1}
SI unit of compressibility
$= m\,kg^{-1}\,s^{2} = m^{2}/N$

reciprocal pascal reciprocal second $Pa^{-1} \cdot s^{-1}$
SI unit of (dynamic) fluidity
$= m\,kg^{-1}\,s = m^{2}/(N \cdot s)$

reciprocal poise P^{-1}
CGS unit of fluidity
10 $Pa^{-1}\cdot s^{-1}$

reciprocal second s^{-1}
SI unit of circular (angular, rotational) frequency
60 min^{-1}
Note: also SI unit of collision rate, damping coefficient, decay constant and velocity gradient; see becquerel, curie, hertz, neper per second, radian per second and rutherford

reciprocal second reciprocal cubic metre $s^{-1}\cdot m^{-3}$
SI unit of slowing-down density and (total) neutron source density

reciprocal second reciprocal kilogram $s^{-1}\cdot kg^{-1}$
=becquerel per kilogram (q.v.)

reciprocal second reciprocal square metre $s^{-1}\cdot m^{-2}$
SI unit of molecule flow rate density and neutron fluence rate
Note: see reciprocal square metre reciprocal second

reciprocal second reciprocal tesla $s^{-1}\cdot T^{-1}$
=ampere square metre per joule second (q.v.)

reciprocal square metre m^{-2}
SI unit of particle fluence

reciprocal square metre reciprocal second $m^{-2}\cdot s^{-1}$
SI unit of particle fluence rate, current density of particles and impingement rate
Note: see reciprocal second reciprocal square metre

register ton (—)
tonneau de jauge
Registertonne
unit of volume ('internal capacity' of a vessel)
100 $ft^3 = 2.831\,685\ m^3$

rem obsol. unit of dose equivalent
10^{-2} Sv
Note: short for *r*öntgen *e*quivalent *m*an

rep obsol. unit of absorbed dose
8.38×10^{-3} Gy
Note: short for *r*öntgen *e*quivalent *p*hysical

rev revolution

revolution (r; rev)
tour (tr)
Umdrehung (U)
unit of plane angle
6.283 19 rad ($=2\pi$)
3.6 $\times$ **10^2** ° (degree)
4.0 $\times$ **10^2** g (grade)

revolution per minute (r/min)
tour par minute (tr/min)
Umdrehung/Minute (U/min)
unit of rotational frequency
1 $\text{min}^{-1} = 1.047\ 20 \times 10^{-1}$ rad/s
Note: symbol tr/mn formerly used in France and rev/min in UK and US

revolution per second (r/s)
tour par seconde (tr/s)
Umdrehung/Sekunde (U/s)
unit of rotational frequency
1 $\text{s}^{-1} = 6.283\ 19$ rad/s
Note: symbol rev/min formerly used in UK and US

reyn =poundal second per square foot (q.v.)

rhe (—)
(a) unit of dynamic fluidity
$\text{cP}^{-1} =$ **10^3** $\text{Pa}^{-1}\cdot\text{s}^{-1}$
(b) unit of kinematic fluidity
$\text{cSt}^{-1} =$ **10^6** s/m^2

right angle . . .⌞)
angle droit
rechter Winkel
unit of plane angle
1.570 80 rad
9.0 $\times$**10** ° (degree) (=π/2)
5.4 $\times$**10**^{3} ′ (minute)
3.24 $\times$**10**^{5} ″ (second)
1.0 $\times$**10**^{2} ^{g} (grade)

rod (—)
depr. unit of length
5.029 2 m = **5.5** yd
Note: also called pole or perch

röntgen R
röntgen
Röntgen
unit of exposure
2.58$\times$**10**$^{-4}$ C/kg

röntgen equivalent man
= rem (q.v.)

röntgen metre squared per curie hour R·m^{2}/(Ci·h)
unit of specific gamma ray constant
1.936 94$\times$10^{-18} C·m^{2}/kg

röntgen per second R/s
unit of exposure rate
2.58$\times$**10**$^{-4}$ C/(kg·s) or A/kg

rood (—)
UK unit of area
1.011 71$\times$10^{3} m^{2}
M/S: Table 2

r.p.h. revolutions per hour (r/h)

r.p.m. revolutions per minute (r/min)

r.p.s. revolutions per second (r/s)

rutherford (Rd)
obsolete unit of activity
10^6 Bq = **1** MBq

S

s second; stat

S siemens; svedberg

sabin (—)
FPS unit of equivalent absorption area
1 ft^2 = 9.290 30 × 10^{-2} m^2

savart (—)
unit of dimensionless quantity: frequency interval
Def: the interval between two frequencies, f_1, f_2, having a ratio
$f_1/f_2 = 10^{1/1000} = 1.002\ 31$
$= 1000 \lg (f_1/f_2)$
$= 3.321\ 93 \times 10^{-3}$ octave [$=(10^{-3}/\lg 2 = 1/301)$]

sb stilb

scfh *c*ubic *f*oot per *h*our at *s*tandard reference conditions

scfm *c*ubic *f*oot per *m*inute at *s*tandard reference conditions

scruple (—)
UK and US unit of mass
1.295 98 × 10^{-3} kg
2.0 × **10** gr
Note: see 'apothecaries' units' and Table 10

sec second (s)

secohm name sometimes used for ohm second (q.v.)

second s

seconde
Sekunde
base SI unit of time
Def: see Appendix 1
$1.666\,67 \times 10^{-2}$ min (=1/60)
$2.777\,78 \times 10^{-4}$ h (=1/3600)
$1.157\,41 \times 10^{-5}$ d (=1/186 400)
Note: also SI unit of duration, period, time interval and time constant, half-life, mean life and specific impulse

second ....″

seconde
Sekunde; Altesekunde
oSI unit of plane angle
$4.848\,14 \times 10^{-6}$ rad
$2.777\,78 \times 10^{-4}$ ° (degree) (=1/3600)
$1.666\,67 \times 10^{-2}$ ′ (minute) (=1/60)
$3.086\,42 \times 10^{-4}$ ^{g} (grade) (=1/3240)
$3.086\,42 \times 10^{-6}$ ∟ (right angle)
Note: the second can be subdivided decimally

second see: centesimal second

second per cubic metre s/m^3

SI unit of resistance (fluid flow)

second per litre s/l, s/L

oSI unit of resistance (fluid flow)
10^3 s/m^3

second per metre squared s/m^2

SI unit of kinematic fluidity

second squared per kilogram s^2/kg

=square metre per joule (q.v.)

Sek. G abbr for *Sekunde*=second (s)

sh cwt short hundredweight

short hundredweight (sh cwt)
US unit of mass
$4.535\,92 \times 10$ kg
1.0 $\times$ **10^2** lb
$8.928\,57 \times 10^{-1}$ hundredweight (UK)
M/S: Table 8

short ton (sh tn)
US unit of mass
$9.071\,85 \times 10^2$ kg (907.184 74 kg)
2.0 $\times$ **10^3** lb
$8.928\,57 \times 10^{-1}$ ton (UK)
M/S: Table 8
Note: in the US usually referred to as ton if there is no danger of confusion with long ton

sh tn short ton

Siegbahn unit see: X-unit

siemens S
siemens
Siemens
SI unit of conductance, admittance, modulus of admittance, susceptance
$= \mathrm{m^{-2}\ kg^{-1}\ s^3\ A^2} = \mathrm{A/V} = \Omega^{-1}$

siemens metre per square millimetre S·m/mm^2
mSI unit of conductivity
10^6 S/m = **1** MS/m

siemens per metre S/m
SI unit of conductivity
$= \mathrm{m^{-3}\ kg^{-1}\ s^3\ A^2}$

siemens square metre per mole S·m^2/mol
SI unit of molar conductivity

sievert Sv
SI unit of dose equivalent
$= \mathrm{m^2\ s^{-2}} = \mathrm{J/kg} = \mathrm{N{\cdot}m/kg}$

sk skot

skot (sk)
obsol. unit of scotopic luminance
$3.183\ 10 \times 10^{-4}$ cd/m$^2 = 10^{-3}$ asb

slug (—)
ft-lbf-s unit of mass
$1.459\ 39 \times 10$ kg
$3.217\ 40 \times 10$ lb (=9.806 65/0.3048)
Note: sometimes called gee pound

slug foot squared (slug·ft^2)
ft-lbf-s unit of moment of inertia
1.355 82 kg·m^2

slug per cubic foot (slug/ft^3)
ft-lbf-s unit of (mass) density
$5.153\ 79 \times 10^2$ kg/m^3

sn sthène

sone (—)
sone
Sone
unit of dimensionless quantity: loudness
$s = 2^{(p-40)/10}$ or approx. $\lg s = 0.03p - 1.2$
where s = number of sones, p = number of phons
Note: 1 on the sone scale corresponds to 40 on the phon scale

sp spat

spat (sp)
(a) unit of solid angle
$1.256\ 64 \times 10$ sr (=4π)
Note: 1 sp = the solid angle of the sphere
(b) obsol. unit of length
10^{12} m = **1** Tm

sq.; sq depr. UK and US abbr. for square (e.g. sq ft = ft^2)

square centimetre cm^2
mSI and CGS unit of area
1.0 $\times$ **10**$^{-4}$ m^2
1.550 00 $\times 10^{-1}$ in^2

square centimetre per dyne cm^2/dyn
CGS unit of compressibility
10 Pa^{-1}

square centimetre per erg cm^2/erg
CGS unit of spectral cross section
10^3 m^2/J

square centimetre per kilogram-force (cm^2/kgf)
unit of compressibility
1.019 72 $\times 10^{-5}$ Pa^{-1}

square centimetre per steradian erg $cm^2/(sr{\cdot}erg)$
CGS unit of spectral angular cross section
10^3 $m^2/(sr{\cdot}J)$

square chain (—)
depr. unit of area
4.046 86 $\times 10^2$ m^2
4.84 $\times$ **10**^{2} yd^2

square degree (□°)
depr. unit of solid angle
3.046 17 $\times 10^{-4}$ sr [$=(\pi/180)^2$]

square foot ft^2
UK and US unit of area
9.290 304 $\times$ **10**$^{-2}$ m^2
M/S: Table 2

square foot hour degree Fahrenheit per British thermal unit foot $ft^2{\cdot}h{\cdot}°F/(Btu{\cdot}ft)$
unit of thermal resistivity
5.777 89 $\times 10^{-1}$ $m{\cdot}K/W$

square foot hour degree Fahrenheit per British thermal unit inch $ft^2 \cdot h \cdot °F/(Btu \cdot in)$
unit of thermal resistivity
6.933 47 m·K/W

square foot per hour ft^2/h
=foot squared per hour (q.v.)

square foot per pound ft^2/lb
FPS unit of specific surface
$2.048\,16 \times 10^{-1}$ m^2/kg

square foot per poundal ft^2/pdl
FPS unit of compressibility
$6.719\,69 \times 10^{-1}$ Pa^{-1}

square foot per pound-force (ft^2/lbf)
ft-lbf-s unit of compressibility
$2.088\,54 \times 10^{-2}$ Pa^{-1}

square foot per second ft^2/s
FPS unit of thermal diffusivity
$\mathbf{9.290\,304 \times 10^{-2}}$ m^2/s

square foot per ton-force ($ft^2/tonf$)
UK unit of compressbility
$9.323\,85 \times 10^{-6}$ Pa^{-1}

square grade ($\square^g$)
depr. unit of solid angle
$2.467\,40 \times 10^{-4}$ sr [$=(\pi/200)^2$]

square inch in^2
UK and US unit of area
$\mathbf{6.451\,6 \times 10^{-4}}$ m^2
6.451 6 cm^2
M/S: Table 2

square inch per pound-force (in^2/lbf)
unit of compressibility
$1.450\,38 \times 10^{-4}$ Pa^{-1}

square inch per ton-force (in^2/tonf)
UK unit of compressibility
$6.474\,90 \times 10^{-8}$ Pa^{-1}

square inch square foot in^2·ft^2
unit of second moment of area
$5.993\,73 \times 10^{-5}$ m^4

square kilometre km^2
mSI unit of area
$\mathbf{10^6}$ m^2 = $\mathbf{10^4}$ a = $\mathbf{10^2}$ ha
$3.861\,02 \times 10^{-1}$ mile^2
$2.471\,05 \times 10^2$ acre

square metre m^2
SI unit of area
$\mathbf{10^{-4}}$ ha = $\mathbf{10^{-2}}$ a = $\mathbf{10^{28}}$ b (barn)
$3.861\,02 \times 10^{-7}$ mile^2
$2.471\,05 \times 10^{-4}$ acre
$1.195\,99$ yd^2
$1.076\,39 \times 10$ ft^2
$1.550\,00 \times 10^3$ in^2
Note: also unit of equivalent absorption area, nuclear quadrupole moment, cross section and atomic attenuation coefficient; cf. metre squared

square metre kelvin per watt m^2·K/W
SI unit of thermal insulance
= kg^{-1} s^3 K

square metre per joule m^2/J
SI unit of spectral cross section
= kg^{-1} s^2
$\mathbf{10^{-3}}$ cm^2/erg = $\mathbf{10^{21}}$ b/erg

square metre per kilogram m^2/kg
SI unit of mass attenuation coefficient, mass energy transfer coefficient, mass absorption coefficient, mass energy absorption coefficient and specific surface
$4.882\,43$ ft^2/lb

square metre per kilogram-force second $m^2/(kgf{\cdot}s)$
m-kgf-s unit of fluidity
$1.019\,72 \times 10^{-1}$ $Pa^{-1}{\cdot}s^{-1}$

square metre per mole m^2/mol
SI unit of molar absorption coefficient and molar attenuation coefficient

square metre per newton m^2/N
=reciprocal pascal (q.v.)

square metre per newton second $m^2/(N{\cdot}s)$
=reciprocal pascal reciprocal second (q.v.)

square metre per second m^2/s
SI unit of diffusion coefficient, thermal diffusion coefficient and thermal diffusivity
=metre squared per second (q.v.)

square metre per steradian m^2/sr
SI unit of angular cross section
$\mathbf{10^{28}}$ b/sr

square metre per steradian joule $m^2/(sr{\cdot}J)$
SI unit of spectral angular cross section
$\mathbf{10^{-3}}$ $cm^2/(sr{\cdot}erg)$
$\mathbf{10^{21}}$ $b/(sr{\cdot}erg)$

square metre per volt second $m^2/(V{\cdot}s)$
SI unit of mobility
$= kg^{-1}\ s^2\ A = m^2/Wb$

square metre per weber m^2/Wb
=square metre per volt second (q.v.)

square micrometre μm^2
mSI unit of area
$\mathbf{10^{-12}}$ m^2
Note: formerly called square micron

square micron (μ^2)
micron carré
Quadratmikron
=square micrometre (q.v.)

square mile $mile^2$
UK and US unit of area
2.589 988 11 $\times$ **10**^{6} m^2
2.589 988 km^2
M/S: Table 2
Note: the symbol mi^2 is used in the US

square mile per ton ($mile^2$/UKton)
UK unit of specific surface
2.549 08 $\times 10^3$ m^2/kg
2.549 08 $\times 10^2$ ha/t

square millimetre mm^2
mSI unit of area
1.0 $\times$ **10**$^{-6}$ m^2
1.550 00 $\times 10^{-3}$ in^2
1.973 53 $\times 10^3$ circular mil

square minute (□′)
depr. unit of solid angle
8.461 59 $\times 10^{-8}$ sr

square second (□″)
depr. unit of solid angle
2.350 44 $\times 10^{-11}$ sr

square yard yd^2
UK and US unit of area
8.361 273 6 $\times$ **10**$^{-1}$ m^2
M/S: Table 2

square yard per ton (yd^2/UKton)
UK unit of specific surface
8.229 22 $\times 10^{-4}$ m^2/kg

sr steradian

st stère

St stokes

standard (—)
unit of volume (for timber only)
4.762 28 m^3 = **1.65** × $\mathbf{10^2}$ ft^3
Note: also known as Petrograd standard

standard . . . see: Part 2

standard atmosphere atm
atmosphère normale
physikalische Atmosphäre; *Normalatmosphäre*
unit of pressure
1.013 25 × $\mathbf{10^5}$ Pa (since 1954)
1.013 25 bar
1.033 23 kgf/cm^2 or kp/cm^2
7.6 × $\mathbf{10^2}$ torr
1.469 59 × 10 lbf/in^2

stat (s)
a prefix denoting a CGSe unit (used in US)

statampere (sA)
CGSe unit of electric current
3.335 64 × 10^{-10} A

statampere centimetre squared (sA·cm^2)
CGSe unit of electromagnetic moment
3.335 64 × 10^{-14} A·m^2

statampere per square centimetre (sA/cm^2)
CGSe unit of current density
3.335 64 × 10^{-6} A/m^2

statcoulomb (sC)
CGSe unit of electric charge
3.335 64 × 10^{-10} C

statcoulomb centimetre (sC·cm)
CGSe unit of electric dipole moment
3.335 64 × 10^{-12} C·m

statcoulomb per cubic centimetre (sC/cm^3)
CGSe unit of volume density of charge
$3.335\,64 \times 10^{-4}$ C/m^3

statcoulomb per square centimetre (sC/cm^2)
CGSe unit of electric polarization and electric flux density
For electric polarization:
$3.335\,64 \times 10^{-6}$ C/m^2
For electric flux density:
$2.654\,42 \times 10^{-7}$ C/m^2

statfarad (sF)
CGSe unit of capacitance
$1.112\,65 \times 10^{-12}$ F

stathenry (sH)
CGSe unit of inductance
$8.987\,55 \times 10^{11}$ H

statmho s℧
=statsiemens (q.v.)

statohm (sΩ)
CGSe unit of resistance
$8.987\,55 \times 10^{11}$ Ω

statohm centimetre (sΩ·cm)
CGSe unit of resistivity
$8.987\,55 \times 10^{9}$ Ω·m

statsiemens (sS)
CGSe unit of conductance
$1.112\,65 \times 10^{-12}$ S

statsiemens per centimetre (sS/cm)
CGSe unit of conductivity
$1.112\,65 \times 10^{-10}$ S/m

stattesla (sT)
=CGSe unit of magnetic flux density (q.v.)

statvolt (sV)
CGSe unit of electric potential
$2.997\,92 \times 10^{2}$ V

statvolt per centimetre (sV/cm)
CGSe unit of electric field strength
$2.997\,92 \times 10^{4}$ V/m

statweber (sWb)
=CGSe unit of magnetic flux (q.v.)

steradian sr
stéradian
Steradiant
supplementary SI unit of solid angle
Def: the steradian is the solid angle which, having its vertex in the centre of a sphere, cuts off an area of the surface of the sphere equal to that of a square having sides of length equal to the radius of the sphere
$7.957\,75 \times 10^{-2}$ spat [$=1/(4\pi)$]
$3.282\,81 \times 10^{3}$ $\square^{\circ}$ (square degree) [$=(180/\pi)^2$]
$4.052\,85 \times 10^{3}$ $\square^{g}$ (square grade) [$=(200/\pi)^2$]

stère (st)
stère
Raummeter (rm); *Ster*
unit of volume (for timber only)
=**1** m^3

sthène (sn)
MTS unit of force
$=t{\cdot}m/s^2$
10^{3} N=**1** kN

sthène per square metre (sn/m^2)
=pièze (q.v.)

stilb (sb)
unit of luminance
10^{4} cd/m^2=**1** cd/cm^2

stokes St
stokes
Stokes
CGS unit of kinematic viscosity
$=cm^2/s$
10^{-4} metre squared per second (q.v.)

stone (—)
UK unit of mass
6.350 29 kg
M/S: Table 7

survey foot (—)
US unit of length
Def: 1 US survey foot = (1200/3937) m
$3.048\,006 \times 10^{-1}$ m
1.000 002 ft
Note: used for Coast and Geodetic surveys within the US

Sv sievert

svedberg (S)
unit of sedimentation coefficient
10^{-13} s = **10**$^{-1}$ ps = **10**^{2} fs

T

t tonne

T tera; tesla

talbot (—)
unit of luminous energy
1 lm·s

techma
G abbreviated name for *technische Masseneinheit* (see metric technical unit of mass)

technical atmosphere at
atmosphère technique
technische Atmosphäre
unit of pressure
$9.806\,65 \times 10^4$ Pa
$9.806\,65 \times 10^{-1}$ bar
1.0 kgf/cm^2 or kp/cm^2
$9.678\,41 \times 10^{-1}$ atm
$1.422\,33 \times 10$ lbf/in^2

telegraph nautical mile (—)
obsol. unit of length
$1.855\,32 \times 10^3$ m
6.087×10^3 ft

tera T
SI prefix denoting $\times 10^{12}$. Examples: terabecquerel (TBq), terahertz (THz), terajoule (TJ), terametre (Tm), teraohm (tΩ), terawatt (TW)

tesla T
tesla
Tesla
SI unit of magnetic flux density and magnetic polarization
$= \text{kg s}^{-2}\ \text{A}^{-1} = \text{Wb/m}^2 = \text{V·s/m}^2$

tesla metre T·m
=weber per metre (q.v.)
Note: not to be confused with terametre (Tm)

tesla square metre T·m^2
=weber (q.v.)

tex tex
unit of linear density (mass per unit length)
10^{-6} kg/m = **1** g/km or mg/m
Note: used for textile filaments

th thermie

therm (—)
UK unit of heat energy
$1.055\,06 \times 10^8$ J $= 105.506$ MJ $= \mathbf{10^5}$ Btu

therm per gallon (UK) therm/UKgal
UK unit of calorific value (volume basis)
$2.320\,80 \times 10^{10}$ J/m^3

thermal . . . the use of this adjective with units of electricity (ampere, coulomb, farad, henry, ohm) to obtain analogical units of heat is deprecated

thermie (th)
obsol. unit of heat energy
4.1855×10^6 J $= 4.1855$ MJ $= \mathbf{10^6}$ cal$_{15}$

thou colloquial name for milli-inch (q.v.); short for one *thou*sandth of an inch

TME G symbol for *technische Masseneinheit* – see: metric technical unit of mass. Very rarely used for 10^{-3} ME i.e. 10^{-3} atomic mass unit (old chemical)

ton (UK) (UKton)
UK unit of mass
$1.016\,047 \times 10^3$ kg
1.016 047 t
1.12 short ton
M/S: Table 7
Note: also called long ton or gross ton to distinguish it from short ton (q.v.)

ton (US) (USton)
= short ton (q.v.)

ton, assay see: assay ton

ton, freight see: freight ton

ton, gross other name for ton (UK) (q.v.)

ton, long other name for ton (UK) (q.v.)

ton, measurement other name for freight ton (q.v.)

ton, metric obsol. name for tonne (q.v.)

ton, net other name for short ton (q.v.)

ton, register see: register ton

ton, shipping other name for freight ton (q.v.)

ton, short see: short ton

ton mile (—)
UK unit of mass carried × distance (traffic factor)
1.635 17 t·km

ton mile per gallon (UK) (UK ton·mile/UKgal)
UK unit of mass carried × distance/volume (traffic factor)
$3.596\ 87 \times 10^{-1}$ t·km/dm^3 or t·km/l or t·km/L

ton of refrigeration (—)
US unit of heat flow rate (refrigerating capacity)
$3.516\ 857 \times 10^3$ W or J/s
1.2 $\times$ **10**^{4} Btu/h
Note: based on short ton

ton (of TNT) (—)
unit of energy ('explosive power')
4.18×10^9 J = 4.18 GJ
Note: the energy released in the explosion of 1 tonne of TNT (trinitrotoluene); the multiples kiloton (= 4.18 TJ) and megaton (= 4.18 PJ) are used in practice for expressing the explosive power of a nuclear weapon

ton per cubic yard (UKton/yd^3)
UK unit of (mass) density
$1.328\ 94 \times 10^3$ kg/m^3
$8.296\ 30 \times 10$ lb/ft^3

ton per hour (UKton/h)
UK unit of mass flow rate
$2.822\,35 \times 10^{-1}$ kg/s
$1.016\,05 \times 10^{3}$ kg/h
$6.222\,22 \times 10^{-1}$ lb/s
2.24 $\times$ **10**^{3} lb/h

ton per mile (UKton/mile)
UK unit of linear density
$6.313\,42 \times 10^{-1}$ kg/m
$4.242\,42 \times 10^{-1}$ lb/ft

ton per square mile (UKton/mile^2)
UK unit of surface density
$3.922\,98 \times 10^{-4}$ kg/m^2
3.922 98 kg/ha
3.5 lb/acre
$8.034\,89 \times 10^{-2}$ lb/1000 ft^2
$1.157\,02 \times 10^{-2}$ oz/yd^2
$1.285\,58 \times 10^{-3}$ oz/ft^2

ton per thousand yards (UKton/1000 yd)
UK unit of linear density
1.111 16 kg/m
$7.466\,67 \times 10^{-1}$ lb/ft

tonf ton-force

ton-force (tonf)
UK unit of force
$9.964\,02 \times 10^{3}$ N
$1.016\,05 \times 10^{3}$ kgf or kp
$7.206\,99 \times 10^{4}$ pdl
2.24 $\times$ **10**^{3} lbf

ton-force foot (tonf·ft)
UK unit of moment of force and torque
$3.037\,03 \times 10^{3}$ N·m
$3.096\,91 \times 10^{2}$ kgf·m or kp·m
2.24 $\times$ **10**^{3} lbf·ft

ton-force per foot (tonf/ft)
UK unit of force per unit length
$3.269\,03 \times 10^4$ N/m

ton-force per square foot ($tonf/ft^2$)
UK unit of pressure
$1.072\,52 \times 10^5$ Pa
1.093 66 kgf/cm^2 or kp/cm^2
2.24 $\times$ **10^3** lbf/ft^2

ton-force per square inch ($tonf/in^2$)
UK unit of pressure
$1.544\,43 \times 10^7$ Pa
$1.574\,88 \times 10^2$ kgf/cm^2 or kp/cm^2
2.24 $\times$ **10^3** lbf/in^2

tonne t
tonne
Tonne
oSI unit and MTS base unit of mass
10^3 kg = **1** Mg (megagram)

tonne kilometre t·km
oSI unit of mass carried $\times$ distance (traffic factor)
$6.115\,58 \times 10^{-1}$ UKton mile

tonne kilometre per litre t·km/l or t·km/L
oSI unit of mass carried $\times$ distance/volume (traffic factor)
2.780 20 UKton·mile/UKgal
2.592 80 sh tn·mile/USgal

tonne metre per second squared $t{\cdot}m/s^2$
= sthène (q.v.)

tonne per cubic metre t/m^3
oSI and MTS unit of (mass) density
10^3 kg/m^3 = **1** Mg/m^3 or kg/dm^3

tonne per hectare t/ha
unit of surface density
10 kilogram per square metre (q.v.)

torr (Torr)
torr
Torr
unit of pressure
$1.333\,22 \times 10^{2}$ Pa (=101 325/760)
$1.333\,22 \times 10^{-3}$ bar
$1.359\,51 \times 10$ kgf/m² or kp/m²
1.000 00 mmHg (conventional)
$1.315\,79 \times 10^{-3}$ atm (=1/760)
$1.933\,68 \times 10^{-2}$ lbf/in²

torr litre per second (Torr·l/s)
unit of leak rate used in vacuum technology
$1.333\,22 \times 10^{-1}$ Pa·m³/s or N·m/s
$\mathbf{1.0 \times 10^{-3}}$ lusec

tr F abbr. for *tour*=revolution (r)

tropical year a; a_{trop}
unit of time
365.242 198 78 d in the year 1900 decreasing at the rate of 0.000 006 14 d per century
Note: this is the year on which the calendar is based. It is the time between two consecutive passages (in the same direction) of the sun through the earth's equatorial plane

troy units Obsol. units of mass used in US and UK and including the following:
1 troy pound (lb tr) =**12** troy ounces
1 troy ounce (oz tr) =**20** pennyweights
1 pennyweight (dwt) =**24** grains
1 grain (gr) (no symb. in US) =**1/480** troy ounce
Note: see Table 9; except for grain and troy ounce (q.v.) not lawful for trade in UK

U

u atomic mass unit (unified); sometimes incorrectly for μ

U G abbr. for *Umdrehung*=revolution (r)

ua; UA microampere (μA)

u.e.m. F abbr. for *unité électromagnétique*
=electromagnetic unit (e.m.u.)

u.e.s. F abbr. for *unité électrostatique*
=electrostatic unit (e.s.u.)

UK when attached to the symbol of a unit indicates a unit used in the United Kingdom the name of which is identical with, or similar to, a name of a US unit, e.g. UKgal

US when attached to the symbol of a unit indicates a unit used in the United States the name of which is identical with, or similar to, a name of a United Kingdom unit, e.g. USgal

US Customary units see: tables Appendix 2

UX see: X-unit

V

V volt

VA volt ampere (V·A)

var var
var
Var
unit of reactive power
1 W

volt V
volt
Volt
SI unit of electric potential, potential difference, electromotive force, thermoelectromotive force and Peltier coefficient
$=m^2\ kg\ s^{-3}\ A^{-1}=W/A=A{\cdot}\Omega$
$=Wb/s=J/C$

volt ampere V·A
unit of apparent power
1 W

volt per ampere V/A
=ohm (q.v.)

volt per kelvin V/K
SI unit of Seebeck coefficient and Thomson coefficient

volt per metre V/m
SI unit of electric field strength
$= \mathrm{m\ kg\ s^{-3}\ A^{-1}}$

volt per mil (V/mil)
unit of electric field strength
$3.937\,01 \times 10^{4}$ V/m

volt second V·s
=weber (q.v.)

volt second metre V·s·m
=weber metre (q.v.)

volt second per ampere V·s/A
=henry (q.v.)

volt second per ampere metre V·s/(A·m)
=henry per metre (q.v.)

volt second per metre V·s/m
=weber per metre (q.v.)

volt second per square metre $\mathrm{V{\cdot}s/m^2}$
=tesla (q.v.)

volt squared per kelvin squared $\mathrm{V^2/K^2}$
SI unit of Lorenz coefficient

W

W watt

watt W
watt
Watt
SI unit of power
$= m^2\ kg\ s^{-3} = J/s = V{\cdot}A$
$1.019\ 72 \times 10^{-1}$ kgf·m/s or kp·m/s
$1.359\ 62 \times 10^{-3}$ metric horsepower
$7.375\ 62 \times 10^{-1}$ ft·lbf/s
$1.341\ 02 \times 10^{-3}$ hp (British horsepower)
Note: also unit of heat flow rate and sound energy flux

watt hour W·h
oSI unit of energy
3.6×10^3 J = **3.6** kJ

watt per ampere squared W/A^2
= ohm (q.v.)

watt per centimetre degree Celsius W/(cm·°C)
$= 10^2$ watt per metre kelvin (q.v.)

watt per cubic foot W/ft^3
unit of heat release rate
$3.531\ 47 \times 10$ W/m^3

watt per cubic metre W/m^3
SI unit of heat release rate
$= m^{-1}\ kg\ s^{-3}$
$8.598\ 45 \times 10^{-1}$ $kcal_{IT}/(m^3{\cdot}h)$
1.0 **× 10** $erg/(cm^3{\cdot}s)$
$9.662\ 11 \times 10^{-2}$ $Btu/(ft^3{\cdot}h)$

watt per foot degree Celsius W/(ft·°C)
unit of thermal conductivity
3.280 84 W/(m·K)

watt per kelvin W/K
SI unit of thermal conductance
$= m^2\ kg\ s^{-3}\ K^{-1}$

watt per kilogram W/kg
SI unit of absorbed dose rate and kerma rate
$= m^2\ s^{-3} = J/(kg{\cdot}s) = Gy/s$
$\mathbf{10^4}$ erg/(g·s) = $\mathbf{10^2}$ rad/s

watt per metre degree Celsius W/(m·°C)
= watt per metre kelvin (q.v.)

watt per metre kelvin W/(m·K)
SI unit of thermal conductivity
$= m\ kg\ s^{-3}\ K^{-1} = W/(m{\cdot}°C)$
$= J{\cdot}m/(m^2{\cdot}s{\cdot}K) = J/(m{\cdot}s{\cdot}K) = J/(m{\cdot}s{\cdot}°C)$
1.0 $\times \mathbf{10^5}$ erg/(cm·s·°C)
$2.388\,46 \times 10^{-3}$ $cal_{IT}/(cm{\cdot}s{\cdot}°C)$
$8.598\,45 \times 10^{-1}$ $kcal_{IT}/(m{\cdot}h{\cdot}°C)$
$1.019\,72 \times 10^{-1}$ kgf/(s·°C) or kp/(s·°C)
$5.777\,89 \times 10^{-1}$ Btu/(ft·h·°F)
6.933 47 Btu·in/(ft²·h·°F)

watt per square centimetre W/cm²
$= 10^4$ watt per square metre (q.v.)

watt per square foot W/ft²
unit of heat flow rate
$1.076\,39 \times 10$ W/m²
9.255 29 $kcal_{IT}/(m^2{\cdot}h)$
3.412 14 Btu/(ft²·h)

watt per square inch W/in²
unit of heat flow rate
$1.550\,00 \times 10^3$ W/m²
$1.332\,76 \times 10^3$ $kcal_{IT}/(m^2{\cdot}h)$
$4.913\,48 \times 10^2$ Btu/(ft²·h)

watt per square metre W/m^2

SI unit of density of heat flow rate

$= kg\ s^{-3} = J/(m^2 \cdot s)$

$9.290\ 30 \times 10^{-2}$ W/ft^2

6.451 6 $\times \mathbf{10^{-4}}$ W/in^2

1.0 $\times \mathbf{10^{3}}$ $erg/(cm^2 \cdot s)$

$2.388\ 46 \times 10^{-5}$ $cal_{IT}/(cm^2 \cdot s)$

$8.598\ 45 \times 10^{-1}$ $kcal_{IT}/(m^2 \cdot h)$

$3.169\ 98 \times 10^{-1}$ $Btu/(ft^2 \cdot h)$

Note: also unit of radiant exitance, irradiance, sound intensity and energy fluence rate and Poynting vector

watt per square metre degree Celsius $W/(m^2 \cdot {}^\circ C)$

= watt per square metre kelvin (q.v.)

watt per square metre kelvin $W/(m^2 \cdot K)$

SI unit of coefficient of heat transfer

$= kg\ s^{-3}\ K^{-1} = J/(m^2 \cdot s \cdot K) = W/(m^2 \cdot {}^\circ C)$

$2.388\ 46 \times 10^{-5}$ $cal_{IT}/(cm^2 \cdot s \cdot {}^\circ C)$

$8.598\ 45 \times 10^{-1}$ $kcal_{IT}/(m^2 \cdot h \cdot {}^\circ C)$

1.0 $\times \mathbf{10^{3}}$ $erg/(cm^2 \cdot s \cdot {}^\circ C)$

$1.019\ 72 \times 10^{-1}$ $kgf/(m \cdot s \cdot {}^\circ C)$ or $kp/(m \cdot s \cdot {}^\circ C)$

$1.761\ 10 \times 10^{-1}$ $Btu/(ft^2 \cdot h \cdot {}^\circ F)$

watt per square metre kelvin to the fourth power $W/(m^2 \cdot K^4)$

SI unit of Stefan-Boltzmann constant

$= kg\ s^{-3}\ K^{-4}$

$\mathbf{10^3}$ $erg/(cm^2 \cdot s \cdot K^4)$

watt per stèradian W/sr

SI unit of radiant intensity

$= m^2\ kg\ s^{-3}\ sr^{-1}$

$\mathbf{10^7}$ $erg/(s \cdot sr)$

watt per steradian square metre $W/(sr \cdot m^2)$

SI unit of radiance

$= kg\ s^{-3}\ sr^{-1}$

$\mathbf{10^3}$ $erg/(s \cdot sr \cdot cm^2)$

watt second $W \cdot s$

= joule (q.v.)

watt square metre W·m²

SI unit of first radiation constant

$= m^4\ kg\ s^{-3}$

Wb weber

weber Wb

weber

Weber

SI unit of magnetic flux and fluxoid quantum

$= m^2\ kg\ s^{-2}\ A^{-1} = V{\cdot}s = J/A = H{\cdot}A = T{\cdot}m^2$

weber metre Wb·m

SI unit of magnetic dipole moment

$= m^3\ kg\ s^{-2}\ A^{-1} = N{\cdot}m^2/A = V{\cdot}s{\cdot}m$

weber per ampere Wb/A

= henry (q.v.)

weber per ampere metre Wb/(A·m)

= henry per metre (q.v.)

weber per metre Wb/m

SI unit of magnetic vector potential

$= m\ kg\ s^{-2}\ A^{-1} = V{\cdot}s/m = T{\cdot}m$

weber per square metre Wb/m^2

= tesla (q.v.)

week (—)

semaine

Woche

unit of time

7 d = **168** h = **10 080** min = **604 800** s

wpm words per minute

wt depr. obsol. abbr. attached to the symbol of a unit of mass when it was used as a unit of '*w*eigh*t*'. Example: kg(wt) for kilogram-force (q.v.)

X

X.E. see: X-unit

X.U. X-unit

X-unit (X.U.)
unité X (*UX*)
X-Einheit (*X.E.*)
depr. unit of wavelength
$1.002\,02 \times 10^{-13}$ m
Note: also called Siegbahn unit; also other values were used

Y

yard yd
UK and US unit of length
Def: the yard equals to 0.9144 metre exactly (from 1959 in US, from 1963 in UK)
M/S: Table 1

yard per pound yd/lb
UK and US unit of specific length
2.015 91 m/kg

yd yard

year a
année
Jahr
unit of time
Note: see Julian year and tropical year

yr year (a)

Miscellaneous

γ gamma

δ dioptre

λ lambda

μ micro; micron

μb microbar

μbar microbar

μHg conventional micron of mercury = conventional micrometre of mercury (μmHg)

μin microinch

μm micrometre

μmHg conventional micrometre of mercury

μs microsecond

Ω ohm

℧ mho

....° degree

....′ minute; foot (ft); centigrade (....^{cg})

....′′ second; inch (in); one hundredth of a centigrade (....^{cc})

□° square degree

□′ square minute

□″ square second

□^{g} square grade

% per cent

‰ per thousand

Part 2

A Dictionary of Quantities and Selected Constants

A

ABSOLUTE ACTIVITY OF SUBSTANCE B λ_B Dim: 1
activité absolue du constituant B
absolute Aktivität eines Stoffes B

ABSOLUTE PERMEABILITY see: permeability

ABSOLUTE PERMITTIVITY see: permittivity

ABSOLUTE PRESSURE p_{abs} Dim: $L^{-1}MT^{-2}$
pression absolue
absoluter Druck
SI unit: pascal Pa
other unit: bar bar

ABSOLUTE TEMPERATURE see: thermodynamic t.

ABSORBED DOSE D Dim: L^2T^{-2}
dose absorbée
Energiedosis
SI unit: gray Gy
other unit: rad rad

ABSORBED DOSE RATE $\dot{D}$ Dim: L^2T^{-3}
débit de dose absorbée
Energiedosisrate, Energiedosisleistung
SI unit: gray per second Gy/s

ABSORPTION CROSS SECTION σ_a, σ_A Dim: L^2
section efficace d'absorption
Absorptionsquerschnitt
SI unit: square metre m^2
other unit: barn b

ACCELERATION see also: angular a., local a., sound particle a.

ACCELERATION a Dim: LT^{-2}
accélération
Beschleunigung
SI unit: metre per second squared m/s^2
Note: also called 'linear acceleration'

ACCELERATION DUE TO GRAVITY see: acceleration of free fall

ACCELERATION OF FREE FALL g Dim: LT^{-2}
accélération due à la pesanteur
Fallbeschleunigung
SI unit: metre per second squared m/s^2
other unit: gal Gal (mGal)
Note: also called 'acceleration due to gravity'; see 'standard acceleration due to gravity'

ACCEPTOR IONIZATION ENERGY E_a Dim: L^2MT^{-2}
énergie d'ionisation d'accepteur
Akzeptorionisierungsenergie
SI unit: joule J
oSI unit: electronvolt eV

ACCEPTOR NUMBER DENSITY n_a, N_a Dim: L^{-3}
nombre volumique d'accepteurs, (*densité d'accepteurs*)
Akzeptoren(an)zahldichte
SI unit: reciprocal cubic metre m^{-3}

ACOUSTIC ... see also: sound ...

ACOUSTIC ABSORPTION COEFFICIENT α, (α_a) Dim: 1
facteur d'absorption acoustique
Schallabsorptionsgrad

ACOUSTIC IMPEDANCE Z_a Dim: $L^{-4}MT^{-1}$
impédance acoustique
akustische Impedanz, Flußimpedanz
SI unit: pascal second per metre cubed $Pa{\cdot}s/m^3$

ACTION (H) Dim: L^2MT^{-1}
action
Wirkung
SI unit: joule second $J{\cdot}s$

ACTIVE POWER P Dim: L^2MT^{-3}

puissance active
Wirkleistung
SI unit: watt W

ACTIVITY see: absolute a., relative a. . . ., specific a.

ACTIVITY A Dim: T^{-1}

activité
Aktivität
SI unit: becquerel Bq
other unit: curie Ci
Note: activity of a radioactive substance; also called ('radioactivity')

ACTIVITY COEFFICIENT OF SOLUTE SUBSTANCE B γ_B Dim: 1

coefficient (facteur) d'activité du soluté B
Aktivitätskoeffizient eines gelösten Stoffes B
Note: particularly in a dilute liquid solution; this quantity is a factor rather than a coefficient

ACTIVITY COEFFICIENT OF SUBSTANCE B Dim: 1
(in a liquid or solid mixture) f_B

coefficient (facteur) d'activité du constituant B (dans un mélange liquide ou solide)
Aktivitätskoeffizient eines Stoffes B (in einem Gemisch fester oder flüssiger Stoffe)
Note: this quantity is a factor rather than a coefficient

ACTIVITY OF SOLUTE SUBSTANCE B a_B, $a_{m,B}$ Dim: 1

activité du soluté B, activité relative du soluté B
Aktivität eines gelösten Stoffes B, relative Aktivität eines gelösten Stoffes B
Note: particularly in a dilute liquid solution; also called 'relative activity of solute substance B'

ACTIVITY OF SOLVENT SUBSTANCE A a_A Dim: 1

activité du solvant A, activité relative du solvant A
Aktivität eines Lösungsmittels A
Note: particularly in a dilute liquid solution; also called 'relative activity of solvent substance A'

ADMITTANCE see also: modulus of a.

ADMITTANCE *Y* Dim: $L^{-2}M^{-1}T^{3}I^{2}$
admittance, (admittance complexe)
Admittanz, (komplexe Admittanz)
SI unit: siemens S
M/S: kS, mS, μS
Note: the reciprocal of impedance; also called ('complex admittance')

AFFINITY (of a chemical reaction) *A* $L^{2}MT^{-2}N^{-1}$
affinité (d'une réaction chimique)
Affinität (einer chemischen Reaktion)
SI unit: joule per mole J/mol

ALFVÉN NUMBER *Al* Dim: 1
nombre d'Alfvén
Alfvén-Zahl

ALPHA DISINTEGRATION ENERGY Q_α Dim: $L^{2}MT^{-2}$
énergie de désintégration alpha
Alpha-Zerfallsenergie; Alpha-Umwandlungsenergie
SI unit: joule J
oSI unit: electronvolt eV
Note: the symbol for 'ground state alpha disintegration energy' is $Q_{\alpha,0}$.

AMBIENT PRESSURE p_{amb} Dim: $L^{-1}MT^{-2}$
pression ambiante
umbegender Druck; umgebender Atmosphärendruck
SI unit: pascal Pa
other unit: bar bar

AMOUNT OF SUBSTANCE *n*, (ν) Dim: N
quantité de matière
Stoffmenge
SI unit: mole mol
M/S: kmol, mmol, μmol
Note: SI base quantity

AMOUNT-OF-SUBSTANCE CONCENTRATION . . . see: concentration . . .

AMOUNT-OF-SUBSTANCE FRACTION see: mole fraction . . .

AMOUNT-OF-SUBSTANCE RATIO see: mole ratio

AMPERAGE obsol. name for *electric current* (q.v.) in amperes

ANERGY Note: same units as energy (q.v.)

ANGLE see: Bragg a., loss a., solid a.

ANGLE, (PLANE ANGLE) $\alpha, \beta, \gamma, \theta, \phi, \ldots$ Dim: 1
angle, (angle plan)
Winkel, (ebener Winkel)
SI unit: radian rad
M/S: mrad, μrad
oSI units: degree °, minute ′, second ″

ANGULAR ACCELERATION α Dim: T^{-2}
accélération angulaire
Winkelbeschleunigung
SI unit: radian per second squared rad/s^2

ANGULAR CROSS SECTION σ_Ω Dim: L^2
section efficace différentielle (angulaire)
raumwinkelbezogener Wirkungsquerschnitt; (Winkelquerschnitt)
SI unit: square metre per steradian m^2/sr
other unit: barn per steradian b/sr

ANGULAR FREQUENCY see: circular frequency

ANGULAR MOMENTUM see: moment of momentum

ANGULAR VELOCITY ω Dim: T^{-1}
vitesse angulaire
Winkelgeschwindigkeit
SI unit: radian per second rad/s
Note: cf. rotational frequency

APPARENT POWER $S, (P_s)$ Dim: L^2MT^{-3}

puissance apparente
Scheinleistung
SI unit: watt W
other unit: voltampere V·A

AREA see: diffusion a., equivalent absorption a., migration a., slowing-down a.

AREA $A, (S)$ Dim: L^2

aire, superficie
Fläche, Flächeninhalt
SI unit: square metre m^2
M/S: km^2, dm^2, cm^2, mm^2
other units: hectare ha, are a

AREA DENSITY obsol. name for 'surface density' (q.v.)

ATOMIC ATTENUATION COEFFICIENT μ_a, μ_{at} Dim: L^2

coefficient d'atténuation atomique
atomarer Schwächungskoeffizient
SI unit: square metre m^2

ATOMIC MASS CONSTANT (UNIFIED) m_u Dim: M

constante (unifiée) de masse atomique
atomare Massenkonstante (vereinheitlichte)
SI unit: kilogram kg
oSI unit: (unified) atomic mass unit u
$m_u = 1.660\,565\,5 \times 10^{-27}$ kg
$= 1$ u

ATOMIC NUMBER Z Dim: 1

nombre atomique, numéro atomique
Ordnungszahl; Atomzahl
Note: preferably called 'proton number' (q.v.) but in the periodic table called traditionally 'atomic number'

ATOMIC WEIGHT

poids atomique

Atomgewicht

Note: obsolete – see: relative atomic mass

ATTENUATION COEFFICIENT α Dim: L^{-1}

affaiblissement linéique (de propagation); (constante d'affaiblissement)

Dämpfunkgskoeffizient, Dämpfunsbelag

SI unit: reciprocal metre m^{-1}

Note: also called 'attenuation constant'; the reciprocal is called 'attenuation length'

ATTENUATION LENGTH l Dim: L

longueur d'atténuation

Dämpfungslänge

SI unit: metre m

AVERAGE ENERGY LOSS PER ION PAIR FORMED W_i Dim: L^2MT^{-2}

perte moyenne d'énergie par paire d'ions formée, (perte moyenne d'énergie par charge élémentaire d'un seul signe formée)

mittlerer Energieverlust (Energieaufwand) je erzeugtes Ionenpaar

SI unit: joule J

oSI unit: electronvolt eV

Note: also called 'average energy loss per elementary charge of one sign produced'

AVERAGE LOGARITHMIC ENERGY DECREMENT ξ Dim: 1

décrément logarithmique moyen de l'énergie, paramètre de ralentissement

mittleres logarithmisches Energiedekrement

AVOGADRO CONSTANT L, N_A Dim: N^{-1}

constante d'Avogadro

Avogadro-Konstante

SI unit: reciprocal mole mol^{-1}

$N_A = 6.022\,045 \times 10^{23}\ mol^{-1}$

B

BETA DISINTEGRATION ENERGY Q_β Dim: L^2MT^{-2}
énergie de désintégration bêta
Beta-Zerfallsenergie; Beta-Umwandlungsenergie
SI unit: joule J
oSI unit: electronvolt eV
Note: the symbol for 'ground state beta disintegration energy' is $Q_{\beta, 0}$

BINDING FRACTION b Dim: 1
facteur de liaison
Bindungsanteil

BIOT NUMBER Bi Dim: 1
nombre de Biot
Biot-Zahl

BODENSTEIN NUMBER see: Péclet n. . . .

BOHR MAGNETON μ_B Dim: L^2I
magnéton de Bohr
Bohr-Magneton
SI unit: ampere square metre $A \cdot m^2$
$\mu_B = 9.274\,078 \times 10^{-24}\ A \cdot m^2$
Note: $A \cdot m^2 = J/T$

BOHR RADIUS a_0 Dim: L
rayon de Bohr
Bohr-Radius
SI unit: metre m
$a_0 = 5.291\,770\,6 \times 10^{-11}$ m

BOLTZMANN CONSTANT k Dim: $L^2MT^{-2}\Theta^{-1}$
constante de Boltzmann
Boltzmann-Konstante
SI unit: joule per kelvin J/K
$k = 1.380\,662 \times 10^{-23}$ J/K

BRAGG ANGLE θ Dim: 1
angle de Bragg
Bragg-Reflexionswinkel
SI unit: radian rad
oSI unit: degree °

BREADTH b Dim: L
largeur
Breite
SI unit: metre m
M/S: km, cm, mm, µm, nm, pm, fm
Note: also called 'width'

BULK COMPRESSIBILITY see: compressibility

BULK MODULUS K Dim: $L^{-1}MT^{-2}$
module de compressibilité volumique
Kompressionsmodul; (Kompressibilitätsmodul)
SI unit: pascal Pa
Note: also called 'modulus of compression'; in BRD N/m^2 is used

BULK STRAIN see: volume strain

BURGERS VECTOR $\boldsymbol{b}$ Dim: L
vecteur de Burgers
Burgers-Vektor
SI unit: metre m
M/S: nm

C

CANONICAL PARTITION FUNCTION Q, Z Dim: 1
fonction de partition canonique
kanonische Zustandssumme

CAPACITANCE *C* Dim: $L^{-2}M^{-1}T^{4}I^{2}$

capacité

elektrische Kapazität

SI unit: farad F

M/S: mF, μF, nF, pF

Note: the reciprocal is called 'elastance'

CAPACITIVITY see: permittivity

CAPACITY see: heat c., molar heat c., specific heat c.

CARRIER LIFE TIME τ, τ_n, τ_p Dim: T

durée de vie de porteur

Trägerlebensdauer

SI unit: second s

CELSIUS TEMPERATURE t, θ Dim: Θ

température Celsius

Celsius-Temperatur

SI unit: degree Celsius °C

Note: cf. 'thermodynamic temperature' and 'temperature difference'

CHARACTERISTIC IMPEDANCE OF A MEDIUM Z_c Dim: $L^{-2}MT^{-1}$

impédance acoustique caractéristique d'un milieu

Schall-Kennimpedanz eines Mediums; (*Wellenwiderstand*)

SI unit: pascal second per metre Pa·s/m

CHARGE see: electric c., elementary c.

CHARGE DENSITY see: volume d.

CHARGE NUMBER OF ION *z* Dim: 1

nombre de charge d'un ion, électrovalence

Ladungszahl eines Ions

CHEMICAL POTENTIAL OF SUBSTANCE B μ_B Dim: $L^{2}MT^{-2}N^{-1}$

potentiel chimique du constituant B

chemisches Potential eines Stoffes B

SI unit: joule per mole J/mol

CHROMATICITY CO-ORDINATES x, y, z Dim: 1
coordonnées trichromatiques
Farbwertanteile, Normfarbwertanteile

CIE SPECTRAL TRISTIMULUS VALUES $\bar{x}(\lambda), \bar{y}(\lambda), \bar{z}(\lambda)$ Dim: 1
fonctions colorimétriques CIE: (formerly: *coefficients de distribution CIE*)
Normspektralwerte
Note: formerly called ('CIE-distribution coefficients')

CIRCULAR FREQUENCY ω Dim: T^{-1}
pulsation
Kreisfrequenz, Winkelfrequenz
SI unit: reciprocal second s^{-1}
or: radian per second rad/s
Note: also called 'angular frequency' and 'pulsatance'

CIRCULAR REPETENCY see: circular wavenumber

CIRCULAR WAVENUMBER $k; q$ Dim: L^{-1}
nombre d'onde angulaire; (*nombre d'onde circulaire*)
Kreiswellenzahl; Kreisrepetenz
SI unit: reciprocal metre m^{-1}
Note: also called 'circular repetency'

COEFFICIENT see: acoustic absorption c., activity c. . . ., atomic attenuation c., attenuation c., coupling c., cubic expansion c., damping c., diffusion c., dissipation c., gyromagnetic c., Hall c., leakage c., linear absorption c., linear attenuation c., linear expansion c., Lorenz c., mass absorption c., mass attenuation c., mass energy attenuation c., mass energy transfer c., molar absorption c., molar attenuation c., osmotic c. . . ., Peltier c., phase c., pressure c., propagation c., recombination c., reflection c., relative pressure c., Seebeck c., surface c. . . ., thermal diffusion c., Thomson c., transmission c.

COEFFICIENT OF . . . see also: . . . coefficient, . . . factor, . . . modulus

COEFFICIENT OF FRICTION μ, (f) Dim: 1
coefficient de frottement, facteur de frottement
Reibungszahl
Note: also called 'friction factor'

COEFFICIENT OF HEAT TRANSFER k, K; h, α Dim: $MT^{-3}\Theta^{-1}$
coefficient de transmission thermique
Wärmedurchgangskoeffizient
SI unit: watt per square metre kelvin $W/(m^2 \cdot K)$
Note: also called 'heat transfer coefficient'; cf. 'surface coefficient of heat transfer'

COEFFICIENT OF THERMAL INSULATION see: thermal insulance

COHERENCE LENGTH ξ Dim: L
longueur de cohérence
Kohärenzlänge, Kohärenzabstand
SI unit: metre m

COLLISION RATE ψ Dim: T^{-1}
taux de collision
Stoßrate
SI unit: reciprocal second s^{-1}

COMPLEX ADMITTANCE see: admittance

COMPLEX IMPEDANCE see: impedance

COMPRESSIBILITY κ Dim: $LM^{-1}T^2$
coefficient de compressibilité (volumique)
Kompressibilität
SI unit: reciprocal pascal Pa^{-1}
Note: also called 'bulk compressibility'; in BRD m^2/N is used

COMPTON WAVELENGTH λ_C Dim: L
longueur d'onde de Compton
Compton-Wellenlänge
SI unit: metre m
For the electron: $\lambda_C = 2.426\,308\,9 \times 10^{-12}$ m
For the neutron: $\lambda_{C,n} = 1.319\,590\,9 \times 10^{-15}$ m
For the proton: $\lambda_{C,p} = 1.321\,409\,9 \times 10^{-15}$ m

CONCENTRATION see also: mass c. . . ., molecular c. . . ., spectral c. . . .

CONCENTRATION OF SUBSTANCE B c_B Dim: $L^{-3}N$
concentration du constituant B, concentration en quantité de matière du constituant B
Konzentration eines Stoffes B; Stoffmengenkonzentration eines Stoffes B
SI unit: mole per cubic metre mol/m³
M/S: mol/dm³, kmol/m³
oSI unit: mole per litre mol/l, mol/L
Note: also called 'amount-of-substance concentration of substance B'; formerly called ('molarity of component B')

CONDUCTANCE (electrical) G Dim: $L^{-2}M^{-1}T^{3}I^{2}$
conductance
elektrischer Leitwert, Konduktanz
SI unit: siemens S
M/S: kS, mS, μS
Note: the reciprocal of conductance is called 'resistance'; both conductance to direct current and conductance as the real part of admittance have the same symbol, dimension and unit

CONDUCTANCE (fluid flow) C, U Dim: $L^{3}T^{-1}$
conductance
Strömungsleitwert
SI unit: cubic metre per second m³/s
oSI unit: litre per second l/s, L/s

CONDUCTANCE see: intrinsic c., molecular c., thermal c.

CONDUCTIVITY see: electrolytic c., thermal c.

CONDUCTIVITY γ, σ Dim: $L^{-3}M^{-1}T^{3}I^{2}$
conductivité
elektrische Leitfähigkeit, Konduktivität
SI unit: siemens per metre S/m
M/S: MS/m, kS/m
Note: the reciprocal of conductivity is called 'resistivity'

CONSTANT (only those having the word 'constant' in their name are listed below) see: atomic mass c., Avogadro c., Boltzmann c., decay c., Dirac c., equilibrium c., Faraday c., fine-structure c., first radiation c., gravitational c., Madelung c., molar gas c., Planck c., reactor time c., Richardson c., Rydberg c., second radiation c., solar c., specific gamma ray c., Stefan-Boltzmann c., time c.

COULOMB MODULUS see: shear m.

COUPING COEFFICIENT k, (κ) Dim: 1
facteur de couplage
Kopplungsgrad

COWLING NUMBER Co Dim: 1
nombre de Cowling
Cowling-Zahl

CROSS SECTION see: absorption c., angular c., fission c., macroscopic c., scattering c., spectral angular c., spectral c., total c., total macroscopic c.

CROSS SECTION σ Dim: L^2
section efficace
Wirkungsquerschnitt
SI unit: square metre m^2
other unit: barn b

CROSS SECTION DENSITY see: macroscopic cross section

CUBIC EXPANSION COEFFICIENT α_V, γ Dim: Θ^{-1}
coefficient de dilatation volumique
(*thermischer*) *Volumenausdehnungskoeffizient*
SI unit: reciprocal kelvin K^{-1}
Note: also called ('volume expansion coefficient')

CURIE TEMPERATURE T_C Dim: Θ
température de Curie
Curie-Temperatur
SI unit: kelvin K

CURRENT DENSITY J, (S) Dim: $L^{-2}I$
densité de courant
elektrische Stromdichte
SI unit: ampere per square metre A/m^2
M/S: MA/m^2, A/mm^2, A/cm^2, kA/m^2

CURRENT DENSITY OF PARTICLES J, (S) Dim: $L^{-2}T^{-1}$
densité de courant de particules
Teilchenstromdichte
SI unit: reciprocal square metre reciprocal second $m^{-2} \cdot s^{-1}$

CURRENT FRACTION . . . see: transport number . . .

CURRENT LINKAGE Θ Dim: I
courant totalisé; solénation—IEC
elektrische Durchflutung
SI unit: ampere A

CYCLOTRON ANGULAR FREQUENCY ω_c Dim: T^{-1}
pulsation cyclotron
Zyklotron-Kreisfrequenz
SI unit: reciprocal second s^{-1}
oSI unit: radian per second rad/s
Note: also called 'cyclotron circular frequency'

CYCLOTRON FREQUENCY ν_c Dim: T^{-1}
fréquence de cyclotron
Zyklotron-Frequenz
SI unit: reciprocal second s^{-1}

D

DAMPED NATURAL FREQUENCY (f_d) Dim: T^{-1}
fréquence propre amortie
Eigenfrequenz bei Dämpfung
SI unit: hertz Hz

DAMPING COEFFICIENT δ Dim: T^{-1}

coefficient d'amortissement
Abklingkoeffizient; (Abklingkonstante)
SI unit: reciprocal second s^{-1}
other units: neper per second Np/s

DEBYE CIRCULAR FREQUENCY ω_D Dim: T^{-1}

pulsation de Debye
Debye-Kreisfrequenz, Debye-Frequenz
SI unit: reciprocal second s^{-1}

DEBYE CIRCULAR WAVENUMBER q_D Dim: L^{-1}

nombre d'onde angulaire de Debye
Debye-Kreiswellenzahl
SI unit: reciprocal metre m^{-1}

DEBYE TEMPERATURE Θ_D Dim: Θ

température de Debye
Debye-Temperatur
SI unit: kelvin K

DEBYE-WALLER FACTOR D Dim: 1

facteur de Debye-Waller
Debye-Waller-Faktor

DECAY CONSTANT λ Dim: T^{-1}

constante de désintégration
Zerfall(s)konstante
SI unit: reciprocal second s^{-1}
Note: also called 'disintegration constant'

DEGREE OF DISSOCIATION α Dim: 1

degré de dissociation; fraction de dissociation; facteur de dissociation
Dissoziationsgrad
Note: also called ('dissociation fraction')

DENSITY see: acceptor number d., current d., diffusion coefficient for . . ., donor number d., electric flux d., electromagnetic energy d., electron number d, energy d., energy flux d., hole number d., intrinsic magnetic flux d., ion d., ion number d., linear current d., linear d., magnetic flux d., mass d., molecule flow d., neutron flux d., neutron number d., number d. . . ., particle flux d., radiant energy d., relative d., slowing-down d., sound energy d., spectral concentration . . ., standard d., surface d., total cross section d., total neutron source d., unitary mass d., volume d. . . .

DENSITY ρ Dim: $L^{-3}M$

masse volumique

Dichte; volumenbezogene Masse

SI unit: kilogram per cubic metre kg/m^3

M/S: Mg/m^3, kg/dm^3, g/cm^3

oSI units: t/m^3, kg/l, kg/L (g/ml, g/mL, g/l, g/L)

Note: also called ('mass density'); the reciprocal is called 'specific volume'

DENSITY OF HEAT FLOW RATE q, ϕ Dim: MT^{-3}

densité de flux thermique

Wärmestromdichte

SI unit: watt per square metre W/m^2

DENSITY OF MOLECULAR FLUX see: molecule flow . . .

DENSITY OF STATES N_E, ρ Dim: $L^{-5}M^{-1}T^2$

densité (électronique) d'état

Zustandsdichte

SI unit: reciprocal joule reciprocal cubic metre

$J^{-1} \cdot m^{-3}$

oSI unit: reciprocal electronvolt reciprocal cubic metre

$eV^{-1} \cdot m^{-3}$

DEPTH (h) Dim: L

profondeur

Tiefe

SI unit: metre m

DIAMETER *d, D* Dim: L
diamètre
Durchmesser
SI unit: metre m
M/S: km, cm, mm, μm, nm, pm, fm

DIFFERENTIAL PRESSURE Δp Dim: $L^{-1}MT^{-2}$
pression différentielle
Differenzdruck
SI unit: pascal Pa
other unit: bar bar

DIFFUSION AREA L^2 Dim: L^2
aire de diffusion
Diffusionsfläche
SI unit: metre squared m^2

DIFFUSION COEFFICIENT *D* Dim: L^2T^{-1}
coefficient de diffusion
Diffusionskoeffizient
SI unit: square metre per second m^2/s

DIFFUSION COEFFICIENT (FOR NEUTRON NUMBER DENSITY) D, D_n Dim: L^2T^{-1}
coefficient de diffusion (*pour le nombre volumique de neutrons*)
Diffusionskoeffizient (*für Neutronen*(*an*)*zahldichte*)
SI unit: square metre per second m^2/s

DIFFUSION COEFFICIENT FOR NEUTRON FLUENCE RATE D_ϕ, (D) Dim: L
coefficient de diffusion pour le débit de fluence de neutrons
Diffusionskoeffizient für Neutronenflußdichte
SI unit: metre m
Note: also called 'diffusion coefficient for neutron flux density'

DIFFUSION LENGTH $L, L_\mathrm{n}, L_\mathrm{p}$ Dim: L
longueur de diffusion
Diffusionslänge
SI unit: metre m

DIRAC CONSTANT $\hbar$ Dim: L^2MT^{-1}

constante réduite de Planck
Dirac-h, h quer
SI unit: joule second J·s
Note: also called '*h*-bar' or '*h*-line' – see 'Planck constant'; only the French name is 'official'

DIRECTIONAL SPECTRAL EMISSIVITY $\varepsilon(\lambda, \theta, \phi)$ Dim: 1

émissivité spectrale directionnelle
gerichteter spektraler Emissionsgrad

DISINTEGRATION CONSTANT see: decay constant

DISPLACEMENT see: electric flux density

DISPLACEMENT VECTOR OF ION $\boldsymbol{u}$ Dim: L

vecteur de déplacement d'un ion
Verschiebungsvektor eines Ions
SI unit: metre m

DISSIPATION COEFFICIENT δ, (ψ) Dim: 1

facteur de dissipation
Schalldissipationsgrad

DISSIPATION FACTOR d Dim: 1

facteur de dissipation (*de pertes*)
Verlustfaktor

DISSOCIATION FRACTION see: degree of dissociation

DONOR IONIZATION ENERGY E_d Dim: L^2MT^{-2}

énergie d'ionisation de donneur
Donatorionisierungsenergie
SI unit: joule J
oSI unit: electronvolt eV

DONOR NUMBER DENSITY n_d, N_d Dim: L^{-3}

nombre volumique de donneurs, (*densité de donneurs*)
Donnatoren(*an*)*zahldichte*
SI unit: reciprocal cubic metre m^{-3}

DOSE see absorbed d., ion d.

DOSE EQUIVALENT *H* Dim: L^2T^{-2}
équivalent de dose
Äquivalentdosis
SI unit: sievert Sv
other unit: rem rem

DURATION *t* Dim: T
durée
Dauer
SI unit: second s
Note: for M/S and oSI units see: time

DYNAMIC MOMENT OF INERTIA see: moment of inertia

DYNAMIC VISCOSITY see: viscosity

E

EFFECTIVE MASS m^* Dim: M
masse effective
effektive Masse
SI unit: kilogram kg

EFFECTIVE MULTIPLICATION FACTOR k_{eff} Dim: 1
facteur effectif de multiplication
Effektivwert des Multiplikationsfaktors

EFFICIENCY η Dim: 1
rendement
Wirkungsgrad

ELECTRIC ... see: the same without 'electric'

ELECTRIC CHARGE *Q* Dim: TI
charge électrique, quantité d'électricité
elektrische Ladung, Elektrizitätsmenge
SI unit: coulomb C
M/S: kC, μC, nC, pC
Note: also called 'quantity of electricity'

ELECTRIC CONSTANT see: permittivity of vacuum

ELECTRIC CURRENT *I* Dim: I
courant électrique, (*intensité de courant électrique*)
elektrische Stromstärke
SI unit: ampere A
M/S: kA, mA, μA, nA, pA
Note: SI base quantity

ELECTRIC DIPOLE MOMENT p, (p_e) Dim: LTI
moment de dipôle électrique
elektrisches Dipolmoment
SI unit: coulomb metre C·m

ELECTRIC DIPOLE MOMENT OF MOLECULE p, μ Dim: LTI
moment de dipôle électrique d'une molécule
elektrisches Dipolmoment eines Moleküls
SI unit: coulomb metre C·m

ELECTRIC FIELD STRENGTH E, (K) Dim: $LMT^{-3}I^{-1}$
champ électrique
elektrische Feldstärke
SI unit: volt per metre V/m
M/S: MV/m, kV/m, V/mm, V/cm, mV/m, μV/m

ELECTRIC FLUX Ψ Dim: TI
flux électrique, (*flux de déplacement*)
elektrischer Fluß, (*Verschiebungsfluß*)
SI unit: coulomb C
M/S: MC, kC, mC
Note: also called ('flux of displacement')

ELECTRIC FLUX DENSITY D Dim: $L^{-2}TI$
induction électrique, déplacement
elektrische Flußdichte, elektrische Verschiebung
SI unit: coulomb per square metre C/m^2
M/S: C/cm^2, kC/m^2, mC/m^2, $\mu C/m^2$
Note: also called 'displacement'

ELECTRIC POLARIZABILITY OF MOLECULE α Dim: $M^{-1}T^4I^2$
polarisation (polarisabilité) électrique d'une molécule
elektrische Polarisierabarkeit eines Moleküls
SI unit: coulomb metre squared per volt $C{\cdot}m^2/V$

ELECTRIC POLARIZATION P, (D_i) Dim: $L^{-2}TI$
polarisation électrique
elektrische Polarisation
SI unit: coulomb per square metre C/m^2
M/S: C/cm^2, kC/m^2, mC/m^2, $\mu C/m^2$

ELECTRIC POTENTIAL V, ϕ Dim: $L^2MT^{-3}I^{-1}$
potential électrique
elektrisches Potential
SI unit: volt V
M/S: MV, kV, mV, μV

ELECTRIC SUSCEPTIBILITY χ, χ_e Dim: 1
susceptibilité électrique
elektrische Suszeptibilität

ELECTROLYTIC CONDUCTIVITY κ, σ Dim: $L^{-3}M^{-1}T^3I^2$
conductivité électrolytique
elektrolytische Leitfähigkeit
SI unit: siemens per metre S/m

ELECTROMAGNETIC ENERGY DENSITY w Dim: $L^{-1}MT^{-2}$
densité d'énergie électromagnétique
elektromagnetische Energiedichte
SI unit: joule per cubic metre J/m^3

ELECTROMAGNETIC MOMENT m Dim: L^2I
moment magnétique; (*moment magnétique ampérien*)
elektromagnetisches Moment, Ampèresches magnetisches Moment
SI unit: ampere metre squared $A{\cdot}m^2$
Note: also called ('magnetic moment')

ELECTROMOTIVE FORCE E Dim: $L^2MT^{-3}I^{-1}$
force électromotrice
elektromotorische Kraft
SI unit: volt V
M/S: MV, kV, mV, μV

ELECTRON NUMBER DENSITY n, n_n, n_- Dim: L^{-3}

nombre volumique électronique, (densité électronique)
Elektronen(an)zahldichte
SI unit: reciprocal cubic metre m^{-3}

ELECTRON RADIUS r_e Dim: L

rayon d'électron
Radius des Elektrons, Elektronradius
SI unit: metre m
$r_e = 2.817\,938\,0 \times 10^{-15}$ m

ELEMENTARY CHARGE e Dim: TI

charge élémentaire
Elementarladung
SI unit: coulomb C
$e = 1.602\,189\,2 \times 10^{-19}$ C

EMISSIVITY ... see: spectral emissivity

EMISSIVITY ε Dim: 1

émissivité
Emissionsgrad

ENERGY see: acceptor ionization e., alpha disintegration e., average loss e., average logarithmic e. ..., beta disintegration e., donor ionization e., Fermi e., gap e., Gibbs free e., Hartree e., Helmholz free e., internal e., kinetic e., maximum beta particle e., molar internal e., potential e., radiant e., reaction e., resonance e., Rydberg e., specific Gibbs free e., specific Helmholz free e., specific internal e.

ENERGY $E, (W)$ Dim: L^2MT^{-2}

énergie
Energie
SI unit: joule J
M/S: EJ, PJ, TJ, GJ, MJ, kJ, mJ
oSI units: electronvolt eV (GeV, MeV, keV)
watt hour W·h (GW·h, MW·h, kW·h)

ENERGY DENSITY w Dim: $L^{-1}MT^{-2}$
énergie volumique
Energiedichte
SI unit: joule per cubic metre J/m^3

ENERGY FLUENCE Ψ Dim: MT^{-2}
fluence énergétique
Energiefluenz
SI unit: joule per square metre J/m^2
oSI unit: electronvolt per square metre eV/m^2

ENERGY FLUENCE RATE ψ Dim: MT^{-3}
débit de fluence énergétique
Energieflußdichte
SI unit: watt per square metre W/m^2
oSI unit: electronvolt per square metre second eV/(m^2·s)
Note: also called 'energy flux density'; cf. 'radiant energy fluence rate'

ENERGY FLUX DENSITY see: energy fluence rate

ENERGY IMPARTED ε Dim: L^2MT^{-2}
énergie communiquée
übertragene Energie, auf das Material übertragene Energie
SI unit: joule J

ENTHALPY H, (I) Dim: L^2MT^{-2}
enthalpie
Enthalpie
SI unit: joule J

ENTROPY S Dim: $L^2MT^{-2}\Theta^{-1}$
entropie
Entropie
SI unit: joule per kelvin J/K
M/S: kJ/K

EQUILIBRIUM CONSTANT K Dim: 1
constante d'équilibre
Gleichgewichtskonstante
Note: there are many equilibrium constants; some of them are not always dimensionless

EQUIVALENT ABSORPTION AREA A Dim: L^2
aire d'absorption équivalente
äquivalente Absorptionsfläche
SI unit: square metre m^2
Note: the rider 'of a surface or object' may be added to the name of this quantity

ETA-FACTOR see: neutron yield per absorption

EULER NUMBER Eu Dim: 1
nombre d'Euler
Euler-Zahl

EXCHANGE INTEGRAL J Dim: L^2MT^{-2}
intégrale d'échange
Austauschintegral
SI unit: joule J
oSI unit: electronvolt eV

EXERGY Note: same units as 'energy' (q.v.)

EXPOSURE X Dim: $M^{-1}TI$
exposition
Exposition (DDR term)
SI unit: coulomb per kilogram C/kg
other unit: röntgen R
Note: a quantity called *Ionendosis* ('ion dose'), symbol: J, SI unit: C/kg, is used in BRD and Austria

EXPOSURE RATE $\dot{X}$ Dim: $M^{-1}I$
débit d'exposition
Expositionsleistung (DDR term)
SI unit: coulomb per kilogram second C/(kg·s)
other unit: röntgen per second R/s
Note: a quantity called *Ionendosisrate* ('ion dose rate'), symbol: $\mathring{J}$, SI unit: A/kg, is used in BRD and Austria

F

FACTOR see also: coefficient

see: friction f., Debye-Waller f., dissipation f., effective multiplication f., fast fission f., *g*-factor, heat transfer f., infinite medium multiplication f., internal conversion f., mass transfer f., multiplication f., power f., spectral absorption f., spectral radiance f., spectral reflection f., spectral transmission f., thermal diffusion f., thermal utilization f.

FARADAY CONSTANT F Dim: TIN^{-1}

constante de Faraday
Faraday -Konstante
SI unit: coulomb per mole C/mol
$F = 9.648\,456 \times 10^4$ C/mol

FAST FISSION FACTOR ε Dim: 1

facteur de fission rapide
Schnellspaltfaktor

FERMI CIRCULAR WAVENUMBER k_F Dim: L^{-1}

nombre d'onde angulaire de Fermi
Fermi-Kreiswellenzahl
SI unit: reciprocal metre m^{-1}

FERMI ENERGY E_F, ε_F Dim: $\mathsf{L}^2\mathsf{MT}^{-2}$

énergie de Fermi
Fermi-Energie
SI unit: joule J
oSI unit: electronvolt eV

FERMI TEMPERATURE T_F Dim: Θ

température de Fermi
Fermi-Temperatur
SI unit: kelvin K

FIELD STRENGTH see: electric f.s., lower critical f.s., magnetic f.s., thermodynamic f.s., upper critical f.s.

FINE-STRUCTURE CONSTANT α Dim: 1
constante de structure fine
Feinstruktur-Konstante (nach Sommerfeld)
$\alpha = 7.297\,350\,6 \times 10^{-3}$
$\alpha^{-1} = 1.370\,360\,4 \times 10^{2}$

FIRST RADIATION CONSTANT c_1 Dim: L^4MT^{-3}
première constante de rayonnement
erste (Plancksche) Strahlungskonstante
SI unit: watt square metre W·m²
$c_1 = 3.741\,832 \times 10^{-16}$ W·m²

FISSION CROSS SECTION σ_f Dim: L^2
section efficace de fission
Spaltungsquerschnitt
SI unit: square metre m²
other unit: barn b

FLUENCE see: energy f., particle f.

FLUIDITY (ϕ) Dim: $LM^{-1}T$
fluidité
Fluidität
SI unit: reciprocal pascal reciprocal second $Pa^{-1} \cdot s^{-1}$
Note: also called 'dynamic fluidity'; it is the reciprocal of (dynamic) viscosity; kinematic fluidity (not included) is the reciprocal of kinematic viscosity (q.v.)

FLUX see: electric f., luminous f., magnetic f., neutron f., radiant energy f., sound energy f.

FLUX OF DISPLACEMENT see: electric flux

FLUXOID QUANTUM Φ_0 Dim: $L^2MT^{-2}I^{-1}$
quantum de flux
Fluxoidquant, Fluxon
SI unit: weber Wb
$\Phi_0 = 2.067\,850\,6 \times 10^{-15}$ Wb

FOCAL LENGTH f Dim: L
distance focale
Brennweite
SI unit: metre m

FORCE see: electromotive f., magnetomotive f., moment of f.

FORCE F Dim: LMT^{-2}
force
Kraft
SI unit: newton N
M/S: MN, kN, mN, μN

FOURIER NUMBER Fo Dim: 1
nombre de Fourier
Fourier-Zahl

FOURIER NUMBER FOR MASS TRANSFER Fo^* Dim: 1
nombre de Fourier pour transfert de masse
Fourier-Zahl der Stoffübertragung

FRACTION see: binding f., dissociation f., mass f. . . ., mole f. . . ., packing f., volume f.

FREQUENCY see: angular f., circular f., cyclotron angular f., cyclotron f., damped natural f., Debye circular f., Larmor angular f., Larmor f., nuclear precession angular f., rotational f., undamped natural f.

FREQUENCY f, ν Dim: T^{-1}
fréquence
Frequenz; Periodenfrequenz
SI unit: hertz Hz
M/S: THz, GHz, MHz, kHz

FREQUENCY INTERVAL (i_f) Dim: 1
intervalle de fréquence
Frequenzintervall
SI unit: (—)
other unit: octave (—)
Note: French standard NF X 02-207 distinguishes between 'musical interval of two frequences' (their ratio) and 'logarithmic interval of two frequences' (logarithm of their ratio), the latter is identical with 'frequency interval'

FRICTION FACTOR see: coefficient of friction

FROUDE NUMBER *Fr* Dim: 1
nombre de Froude
Froude-Zahl
Note: also called Reech number

FUGACITY OF SUBSTANCE B (IN A GASEOUS MIXTURE) f_B, $\tilde{p}_B$ Dim: $L^{-1} MT^{-2}$
fugacité du constituant B (dans un mélange gazeux)
Fugazität eines Stoffes B (in einem Gasgemisch)
SI unit: pascal Pa

FUNCTION see: canonical partition f., Gibbs free energy, grand-canonical partition f., Helmholz free energy, Massieu f., microcanonical partition f., molecular partition f., Planck f., specific Gibbs free energy, specific Helmholz free energy, work f.

G

GAS CONSTANT see: molar g.c.

GAUGE PRESSURE p_e Dim: $L^{-1}MT^{-2}$
pression effective
atmosphärische Druckdifferenz, Überdruck
SI unit: pascal Pa
other unit: bar bar
Note: the term *Überdruck* is used both for a positive and a negative difference – the term *Unterdruck* is no longer used for the latter

g-FACTOR OF ATOM OR ELECTRON *g* Dim: 1
facteur g d'un atome ou d'un électron; (*valeur g . . .*)
g-Faktor eines Atoms oder eines Elektrons
Note: also called ('*g*-value of . . .')

g-FACTOR OF NUCLEUS OR NUCLEAR PARTICLE g Dim: 1
facteur g d'un noyau ou d'une particule nucléaire; (*valeur g d'...*)
g-Faktor eines Kernes oder eines Teilchens
Note: also called ('*g*-value of . . .')

GIBBS FREE ENERGY G Dim: L^2MT^{-2}
enthalpie libre
freie Enthalpie, Gibbs-Funktion
SI unit: joule J
Note: also called 'Gibbs function'

GRAND-CANONICAL PARTITION FUNCTION Ξ Dim: 1
fonction de partition grand-canonique
großkanonische Zustandssumme
Note: also called 'grand partition function'

GRASHOF NUMBER Gr Dim: 1
nombre de Grashof
Grashof-Zahl

GRASHOF NUMBER FOR MASS TRANSFER Gr^* Dim: 1
nombre de Grashof pour transfert de masse
Grashof-Zahl der Stoffübertragung

GRAVITATIONAL CONSTANT G, (f) Dim: $L^3M^{-1}T^{-2}$
constante de gravitation
Gravitationskonstante
SI unit: newton square metre per kilogram squared $N{\cdot}m^2/kg^2$
$G = 6.6720 \times 10^{-11}\ N{\cdot}m^2/kg^2$

GROWTH COEFFICIENT σ Dim: T^{-1}
coefficient d'accroissement
Anklingkoeffizient, Wuchskoeffizient
SI unit: reciprocal second s^{-1}

GRÜNEISEN PARAMETER γ, Γ Dim: 1
paramètre de Grüneisen
Grüneisen-Parameter

g-VALUE OF . . . see: *g*-factor of . . .

GYROMAGNETIC COEFFICIENT γ Dim: $M^{-1}TI$

coefficient gyromagnétique, rapport gyromagnétique
gyromagnetischer Koeffizient
SI unit: ampere square metre per joule second $A \cdot m^2/(J \cdot s)$
Note: also called ('gyromagnetic ratio')
G.c. of the proton: $\gamma_p = 2.675\,198\,7 \times 10^8$ $A \cdot m^2/(J \cdot s)$

H

HALF-LIFE $T_{1/2}$ Dim: T

période radioactive
Halbwertzeit
SI unit: second s

HALF-THICKNESS $d_{1/2}$ Dim: L

couche de demi-atténuation
Halbwertschicht, Halbwertdicke
SI unit: metre m
Note: also called 'half-value thickness'

HALL COEFFICIENT A_H, R_H Dim: $L^3T^{-1}I^{-1}$

coefficient de Hall
Hall-Koeffizient
SI unit: cubic metre per coulomb m^3/C

HARTMANN NUMBER *Ha* Dim: 1

nombre de Hartmann
Hartmann-Zahl

HARTREE ENERGY E_h Dim: L^2MT^{-2}

énergie de Hartree
Hartree-Energie
SI unit: joule J
$E_h = 4.359\,81 \times 10^{-18}$ J
Note: $E_h = 2\ Ry$ (Ry = Rydberg energy, q.v.)

h-BAR see: Dirac constant

HEAT... see also: thermal...

HEAT Q Dim: L^2MT^{-2}
quantité de chaleur
Wärme, Wärmemenge
SI unit: joule J
M/S: EJ, PJ, TJ, GJ, MJ, kJ, mJ
Note: also called 'quantity of heat'

HEAT CAPACITY C Dim: $L^2MT^{-2}\Theta^{-1}$
capacité thermique
Wärmekapazität
SI unit: joule per kelvin J/K
M/S: kJ/K

HEAT FLOW RATE Φ Dim: L^2MT^{-3}
flux thermique
Wärmestrom
SI unit: watt W
M/S: kW

HEAT TRANSFER COEFFICIENT see: coefficient of heat transfer

HEAT TRANSFER FACTOR j Dim: 1
facteur de transfert de chaleur
Wärmeübertragungsfaktor

HEIGHT h Dim: L
hauteur
Höhe
SI unit: metre m
M/S: km, cm, mm, µm, nm, pm, fm

HELMHOLTZ FREE ENERGY A, F Dim: L^2MT^{-2}
énergie libre
freie Energie, Helmholtz-Funktion
SI unit: joule J
Note: also called 'Helmholtz function'

H-LINE see: Dirac constant

HOLE NUMBER DENSITY p, n_p, n_+ Dim: L^{-3}
nombre volumique de trous, (densité de trous)
Löcher(an)zahldichte
SI unit: reciprocal cubic metre m^{-3}

HYPERFINE STRUCTURE QUANTUM NUMBER F Dim: 1
nombre quantique de structure hyperfine
Hyperfeinstruktur-Quantenzahl

I

ILLUMINANCE $E, (E_v)$ Dim: $L^{-2}J$
éclairement lumineux, éclairement
Beleuchtungsstärke
SI unit: lux lx
Note: formerly called ('illumination')

IMPEDANCE see: acoustic i., characteristic i. . . ., complex i., mechanical i., modulus of i., specific acoustic i.

IMPEDANCE Z Dim: $L^2MT^{-3}I^{-2}$
impédance, (impédance complexe)
Impedanz, (komplexe Impedanz)
SI unit: ohm Ω
M/S: MΩ, kΩ, mΩ
Note: the reciprocal of 'admittance'; also called ('complex impedance')

IMPINGEMENT RATE v Dim: $L^{-2}T^{-1}$
taux (surfacique) d'incidence
flächenbezogene Stoßrate, Flächenstoßrate
SI unit: reciprocal square metre reciprocal second
$m^{-2} \cdot s^{-1}$

INDUCTANCE see: mutual i., self i.

INFINITE MEDIUM MULTIPLICATION FACTOR k_∞ Dim: 1
facteur infini de multiplication
Multiplikationsfaktor bei unendlich ausgedehntem Medium

INTENSITY see: luminous i., radiant i., sound i.

INTERNAL CONVERSION FACTOR α Dim: 1
facteur de conversion interne
Konversionskoeffizient; Koeffizient der inneren Konversion

INTERNAL ENERGY U, (E) Dim: L^2MT^{-2}
énergie interne
innere Energie
SI unit: joule J

INTRINSIC CONDUCTANCE C_i, U_i Dim: L^3T^{-1}
conductance intrinsèque
Strömungseigenleitwert, charakteristischer Strömungsleitwert
SI unit: cubic metre per second m^3/s
oSI unit: litre per second l/s, L/s

INTRINSIC MAGNETIC FLUX DENSITY see: magnetic polarization

ION DENSITY see: ion number density

ION DOSE J Dim: $M^{-1}TI$
dose ionique
Ionendosis
SI unit: coulomb per kilogram C/kg
Note: the English and French terms are mere translations of the German term; the quantity used in English and French speaking countries is 'exposure', symbol X, the same unit

ION DOSE RATE $\dot{J}$ Dim: $M^{-1}I$

débit de dose ionique
Ionendosisrate, Ionendosisleistung
SI unit: ampere per kilogram A/kg
Note: the English and French terms are mere translations of the German term; the quantity used in English and French speaking countries is 'exposure rate', symbol $\dot{X}$, SI unit C/(kg·s)

ION NUMBER DENSITY n^{+}, n^{-} Dim: L^{-3}

nombre volumique d'ions
Ionen(an)zahldichte
SI unit: reciprocal cubic metre m^{-3}
Note: also called 'ion density'

IONIC STRENGTH I Dim: $M^{-1}N$

force ionique
Ionenestärke
SI unit: mole per kilogram mol/kg

IRRADIANCE E, (E_e) Dim: MT^{-3}

éclairement énergétique
Bestrahlungsstärke
SI unit: watt per square metre W/m^2

ISENTROPIC EXPONENT κ Dim: 1

exposant isentropique
Isentropenexponent

ISOTOPIC NUMBER see: neutron excess number

J

JERK (r), (h) Dim: LT^{-3}

saccade
Ruck
SI unit: metre per second cubed m/s^3

K

KERMA K Dim: L^2T^{-2}
kerma
Kerma
SI unit: gray Gy
Note: short for *k*inetic *e*nergy *r*eleased in *ma*tter

KERMA RATE $\dot{K}$ Dim: L^2T^{-3}
débit de kerma
Kermarate, Kermaleistung
SI unit: gray per second Gy/s

KINEMATIC VISCOSITY ν Dim: L^2T^{-1}
viscosité cinématique
kinematische Viskosität
SI unit: metre squared per second m^2/s
M/S: mm^2/s

KINETIC ENERGY E_k, K, T Dim: L^2MT^{-2}
énergie cinétique
kinetische Energie
SI unit: joule J
oSI unit: electronvolt eV
Note: cf. 'energy'

KNUDSEN NUMBER Kn Dim: 1
nombre de Knudsen
Knudsen-Zahl

L

LANDAU-GINZBURG PARAMETER κ Dim: 1
paramètre de Landau-Ginzburg
Landau-Ginzburg-Parameter

LARMOR ANGULAR FREQUENCY ω_L Dim: T^{-1}
pulsation de Larmor
Larmor-Kreisfrequenz
SI unit: reciprocal second s^{-1}
oSI unit: radian per second rad/s
Note: also called 'Larmor circular frequency'

LARMOR FREQUENCY ν_L Dim: T^{-1}
fréquence de précession de Larmor
Larmor-Frequenz
SI unit: reciprocal second s^{-1}

LATENT HEAT L Dim: L^2MT^{-2}
chaleur latente
latente Wärme(menge), Umwandlungswärme
SI unit: joule J

LATTICE VECTOR $\boldsymbol{R}, \boldsymbol{R}_0, \boldsymbol{T}$ Dim: L
vecteur de réseau
Gittervektor
SI unit: metre m
M/S: nm

LEAK RATE Q_L Dim: L^2MT^{-3}
débit de fuite; taux de fuite
Leckrate
SI unit: pascal cubic metre per second $Pa\cdot m^3/s$
Note: used in vacuum technology

LEAKAGE COEFFICIENT σ Dim: 1
facteur de dispersion
Streufaktor, Streugrad

LENGTH see: attenuation l., coherence l., diffusion l., focal l., migration l., slowing-down l. See also: wavelength

LENGTH l, (L) Dim: L
longueur
Länge
SI unit: metre m
M/S: km, cm, mm, μm, nm, pm, fm
oSI units: astronomic(al) unit (AU), parsec pc
Note: SI base quantity

LENGTH OF PATH s Dim: L
longueur curviligne
Weglänge, Kurvenlänge
SI unit: metre m
M/S: km, cm, mm, µm, nm, pm, fm

LETHARGY u Dim: 1
léthargie
Lethargie

LEVEL see: loudness l., sound intensity l., sound power l., sound pressure l.

LEVEL WIDTH Γ Dim: L^2MT^{-2}
largeur de niveau
Niveaubreite
SI unit: joule J
oSI unit: electronvolt eV

LEWIS NUMBER Le Dim: 1
nombre de Lewis
Lewis-Zahl

LIGHT EXPOSURE H Dim: $L^{-2}TJ$
exposition lumineuse; (formerly: *quantité d'éclairement*)
Belichtung
SI unit: lux second lx·s
oSI unit: lux hour lx·h
Note: formerly called ('quantity of illumination')

LINEAR ABSORPTION COEFFICIENT a Dim: L^{-1}
coefficient d'absorption linéique
(*linearer*) *Absorptionskoeffizient*
SI unit: reciprocal metre m^{-1}

LINEAR ACCELERATION see: acceleration

LINEAR ATTENUATION COEFFICIENT μ, μ_l Dim: L^{-1}
coefficient d'atténuation linéique
linearer Schwächungskoeffizient
SI unit: reciprocal metre m^{-1}
Note: also called 'linear extinction coefficient'

LINEAR CURRENT DENSITY A, (α) Dim: $L^{-1}I$
densité linéique de courant
elektrischer Strombelag
SI unit: ampere per metre A/m
M/S: kA/m, A/mm, A/cm

LINEAR DENSITY ρ_l Dim: $L^{-1}M$
masse linéique
längenbezogene Masse, Massenbelag, Massenbehang
SI unit: kilogram per metre kg/m
M/S: mg/m
Note: formerly called ('mass per unit length'); the reciprocal is called 'specific length'

LINEAR ENERGY TRANSFER L Dim: LMT^{-2}
transfert linéique d'énergie
lineares Energieübertragungsvermögen
SI unit: joule per metre J/m
oSI unit: electronvolt per metre eV/m
Note: sometimes abbreviated LET

LINEAR EXPANSION COEFFICIENT α_l Dim: Θ^{-1}
coefficient de dilatation linéique
(*thermischer*) *Längenausdehnungskoeffizient*
SI unit: reciprocal kelvin K^{-1}
Note: also called 'thermal coefficient of linear expansion'

LINEAR EXTINCTION COEFFICIENT see: linear attenuation coefficient

LINEAR IONIZATION BY A PARTICLE N_{il} Dim: L^{-1}
ionisation linéique d'une particule
lineare Ionisation eines Teilchens
SI unit: reciprocal metre m^{-1}

LINEAR STRAIN e, ε Dim: 1
dilatation linéique relative
Dehnung, relative Längenänderung
Note: also called ('relative elongation')

LINEAR VELOCITY see: velocity

LOCAL ACCELERATION OF FREE FALL g Dim: LT^{-2}
accélération locale en chute libre
örtliche Fallbeschleunigung
SI unit: metre per second squared m/s^2
Note: the same as 'acceleration of free fall'

LOGARITHMIC DECREMENT Λ Dim: 1
décrément logarithmique
logarithmisches Dekrement
SI unit: —
other units: neper Np

LONDON PENETRATION DEPTH λ_L Dim: L
profondeur de pénétration de London
London-Eindring(ungs)tiefe
SI unit: metre m

LONG RANGE ORDER PARAMETER s Dim: 1
paramètre d'ordre à grande distance
Fernordnungsparameter

LORENZ COEFFICIENT L Dim: $L^4M^2T^{-6}I^{-2}\Theta^{-2}$
coefficient de Lorenz
Lorenz-Koeffizient
SI unit: volt squared per kelvin squared V^2/K^2

LOSCHMIDT CONSTANT N_L Dim: L^{-3}
constante de Loschmidt
Loschmidt-Konstante
SI unit: reciprocal cubic metre m^{-3}
$N_L = N_A/V_{m,0}$
$= 2.686\,75 \times 10^{25}\ m^{-3}$

LOSS ANGLE δ Dim: 1
angle de pertes
Verlustwinkel
SI unit: radian rad

LOUDNESS N Dim: 1
sonie
Lautheit
SI unit: —
other unit: sone (—)

LOUDNESS LEVEL L_N Dim: 1
niveau d'isosonie
Lautstärkepegel
SI unit: —
other unit: phon (—)

LOWER CRITICAL FIELD STRENGTH H_{c1} Dim: $L^{-1}I$
intensité du champ critique inférieur
untere kritische Feldstärke
SI unit: ampere per metre A/m

LUMINANCE L, (L_v) Dim: $L^{-2}J$
luminance
Leuchtdichte
SI unit: candela per square metre cd/m^2

LUMINOUS ... see also: spectral luminous ...

LUMINOUS EFFICACY K Dim: $L^{-2}M^{-1}T^3J$
efficacité lumineuse
photometrisches Strahlungsäquivalent
SI unit: lumen per watt lm/W

LUMINOUS EFFICIENCY V Dim: 1
efficacité lumineuse relative
Hellempfindlichkeitsgrad

LUMINOUS EXITANCE M, (M_v) Dim: $L^{-2}J$
exitance lumineuse; (formerly: *émittance lumineuse*)
spezifische Lichtausstrahlung
SI unit: lumen per square metre lm/m^2
Note: formerly called ('luminous emitance')

LUMINOUS FLUX Φ, (Φ_v) Dim: J
flux lumineux
Lichtstrom
SI unit: lumen lm

LUMINOUS INTENSITY I, (I_v) Dim: J
intensité lumineuse
Lichtstärke
SI unit: candela cd
Note: SI base quantity

M

MACH NUMBER *Ma* Dim: 1
nombre de Mach
Mach-Zahl

MACROSCOPIC CROSS SECTION Σ Dim: L^{-1}
section efficace macroscopique, section efficace volumique
(*volumenbezogener*) *makroskopischer Wirkungsquerschnitt, Wirkungsquerschnittsdichte*
SI unit: reciprocal metre m^{-1}
Note: also called 'cross section density'

MADELUNG CONSTANT α Dim: 1
constante de Madelung, facteur de Madelung
Madelung-Konstante

MAGNETIC CONSTANT see: permeability of vacuum

MAGNETIC DIPOLE MOMENT j Dim: $L^3MT^{-2}I^{-1}$
moment de dipôle magnétique (*coulombien*)
magnetisches Dipolmoment
SI unit: weber metre Wb·m
Note: the unit newton metre squared per ampere $N{\cdot}m^2/A$ is sometimes used for this quantity

MAGNETIC FIELD STRENGTH H Dim: $L^{-1}I$
champ magnétique
magnetische Feldstärke, magnetische Erregung
SI unit: ampere per metre A/m
M/S: kA/m, A/mm, A/cm

MAGNETIC FLUX Φ Dim: $L^2MT^{-2}I^{-1}$
flux magnétique, flux d'induction magnétique
magnetischer Fluß
SI unit: weber Wb
M/S: mWb

MAGNETIC FLUX DENSITY B Dim: $MT^{-2}I^{-1}$
induction magnétique, densité de flux magnétique
magnetische Flußdichte, magnetische Induktion
SI unit: tesla T
M/S: mT, μT, nT
Note: also called 'magnetic induction'

MAGNETIC INDUCTION see: magnetic flux density

MAGNETIC INTENSITY obsol. name for 'magnetic field strengths'

MAGNETIC MOMENT see: electromagnetic moment

MAGNETIC MOMENT OF A PARTICLE (OR NUCLEUS) μ Dim: L^2I
moment magnétique d'une particule (*ou d'un noyau*)
magnetisches Moment eines Teilchens (*oder eines Kernes*)
SI unit: ampere square metre A·m²
Note: see also 'electromagnetic moment'

MAGNETIC POLARIZATION B_i, J Dim: $MT^{-2}I^{-1}$
polarisation magnétique; induction intrinsèque – IEC
magnetische Polarisation
SI unit: tesla T
M/S: mT
Note: also called 'intrinsic magnetic flux density' – IEC

MAGNETIC POTENTIAL DIFFERENCE U_m; U Dim: I
différence de potentiel magnétique; tension magnétique – IEC
magnetische Spannung
SI unit: ampere A
M/S: kA, mA

MAGNETIC QUANTUM NUMBER m_i, M Dim: 1
nombre quantique magnétique
magnetische Quantenzahl

MAGNETIC REYNOLDS NUMBER Rm Dim: 1
nombre de Reynolds magnétique
magnetische Reynolds-Zahl

MAGNETIC SUSCEPTIBILITY κ, (χ_m) Dim: 1
susceptibilité magnétique
magnetische Suszeptibilität

MAGNETIC VECTOR POTENTIAL A Dim: $LMT^{-2}I^{-1}$
potentiel vecteur magnétique
magnetisches Vektorpotential
SI unit: weber per metre Wb/m
M/S: kWb/m, Wb/mm

MAGNETIZATION H_i, M Dim: $L^{-1}I$
aimantation
Magnetisierung
SI unit: ampere per metre A/m
M/S: kA/m, A/mm

MAGNETOMOTIVE FORCE F, F_m Dim: I
force magnétomotrice
magnetomotorische Kraft
SI unit: ampere A

MAGNETON see: Bohr m., nuclear m.

MARGOULIS NUMBER Ms Dim: 1
nombre de Margoulis
Margoulis-Zahl
Note: alternative name for 'Stanton number'

MASS see: effective m., molar m., relative atomic m., relative molecular m., rest m.

MASS m Dim: M
masse
Masse
SI unit: kilogram kg
M/S: Mg, g, mg, μg
oSI unit: tonne t
Note: SI base quantity

MASS ABSORPTION COEFFICIENT a_m, a/ρ Dim: L^2M^{-1}
coefficient d'absorption massique
Massenabsorptionskoeffizient
SI unit: square metre per kilogram m^2/kg

MASS ATTENUATION COEFFICIENT μ_m, μ/ρ Dim: L^2M^{-1}
coefficient d'atténuation massique
Massenschwächungskoeffizient
SI unit: square metre per kilogram m^2/kg

MASS CONCENTRATION OF SUBSTANCE B ρ_B Dim: $L^{-3}M$
concentration en masse du constituant B
Massenkonzentration eines Stoffes B; Partialdichte eines Stoffes B
SI unit: kilogram per cubic metre kg/m^3
oSI unit: kilogram per litre kg/l, kg/L

MASS DEFECT B Dim: M
défaut de masse
Massendefekt
SI unit: kilogram kg
oSI unit: (unified) atomic mass unit u

MASS DENSITY see: density

MASS ENERGY ABSORPTION COEFFICIENT μ_{en}/ρ Dim: L^2M^{-1}
coefficient d'absorption d'énergie massique
Massenenergieabsorptionskoeffizient
SI unit: square metre per kilogram m^2/kg

MASS ENERGY TRANSFER COEFFICIENT μ_{tr}/ρ Dim: L^2M^{-1}
coefficient de transfert d'énergie massique
Massenenergieumwandlungskoeffizient; Massenenergieübertragungskoeffizient
SI unit: square metre per kilogram m^2/kg

MASS EXCESS Δ Dim: M
excès de masse
Massenüberschuß
SI unit: kilogram kg
oSI unit: (unified) atomic mass unit u

MASS FLOW RATE q_m Dim: MT^{-1}
débit-masse
Massenstrom, Massendurchfluß, Massendurchsatz
SI unit: kilogram per second kg/s

MASS FRACTION OF SUBSTANCE B w_B Dim: 1
fraction massique du constituant B
Massenanteil (*Massengehalt*) *eines Stoffes B;* (*Massenbruch eines Stoffes B*)
Note: usually expressed in %, ‰, or ppm; mass fraction in % is intended to replace incorrect 'mass %' and deprecated '% by weight' and 'weight %'

MASS NUMBER see: nucleon number

MASS OF ATOM (OF A NUCLIDE X) m_a, m(X) Dim: M
masse atomique (*d'un nucléide X*), *masse nucléidique*
Nuklidmasse; Masse eines Atoms eines Nuklids; Atommasse
SI unit: kilogram kg
oSI unit: (unified) atomic mass unit u
Note: also called 'nuclidic mass'

MASS OF MOLECULE m Dim: M
masse de la molécule
Masse eines Moleküls
SI unit: kilogram kg

MASS PER UNIT AREA see: surface density

MASS PER UNIT LENGTH see: linear density

MASS PERCENT (mass %)
pour cent en masse (*% en masse*)
Masseprozent (*Masse-%*)
Note: also called 'percent by mass' (% by mass); see 'mass fraction'

MASS TRANSFER FACTOR j_m Dim: 1
facteur de transfert de masse
Massenübertragungsfaktor

MASSIEU FUNCTION J Dim: $L^2MT^{-2}\Theta^{-1}$
fonction de Massieu
Massieu-Funktion
SI unit: joule per kelvin J/K

MAXIMUM BETA PARTICLE ENERGY E_β Dim: L^2MT^{-2}
énergie bêta maximale
Beta-Maximalenergie
SI unit: joule J
oSI unit: electronvolt eV

MAXIMUM SPECTRAL LUMINOUS EFFICACY K_m Dim: $L^{-2}M^{-1}T^3J$
efficacité lumineuse spectrale maximale
Maximalwert des spektralen photometrischen Strahlungsäquivalentes
SI unit: lumen per watt lm/W

MEAN FREE PATH l, λ Dim: L
libre parcours moyen
mittlere freie Weglänge
SI unit: metre m

MEAN FREE PATH OF ELECTRONS l, l_e Dim: L
libre parcours moyen des électrons
mittlere freie Weglänge der Elektronen
SI unit: metre m

MEAN FREE PATH OF PHONONS l_{ph}, Λ Dim: L
libre parcours moyen des phonons
mittlere freie Weglänge der Phononen
SI unit: metre m

MEAN LIFE τ Dim: T
vie moyenne
mittlere Lebensdauer
SI unit: second s

MEAN LINEAR RANGE R, R_l Dim: L
parcours moyen linéaire
mittlere lineare Reichweite
SI unit: metre m

MEAN MASS RANGE R_ρ, (R_m) Dim: $L^{-2}M$
parcours moyen en masse
mittlere Massenreichweite
SI unit: kilogram per square metre kg/m^2

MECHANICAL IMPEDANCE Z_m Dim: MT^{-1}
impédance mécanique
mechanische Impedanz
SI unit: newton second per metre N·s/m

MICROCANONICAL PARTITION FUNCTION Ω Dim: 1
fonction de partition microcanonique
mikrokanonische Zustandssumme

MIGRATION AREA M^2 Dim: L^2
aire de migration
Wanderfläche
SI unit: metre squared m^2

MIGRATION LENGTH M Dim: L
longueur de migration
Wanderlänge
SI unit: metre m

MOBILITY μ Dim: $M^{-1}T^2I$
mobilité
Beweglichkeit
SI unit: square metre per volt second $m^2/(V \cdot s)$

MOBILITY RATIO b Dim: 1
rapport de mobilité
Beweglichkeitsverhältnis

MODULUS see: bulk m., section m., shear m.

MODULUS OF ADMITTANCE $|Y|$ Dim: $L^{-2}M^{-1}T^{3}I^{2}$
module de l'admittance, (*admittance*)
Scheinleitwert, Betrag der Admittanz
SI unit: siemens S
M/S: kS, mS, μS
Note: also called ('admittance')

MODULUS OF COMPRESSION see: bulk modulus

MODULUS OF ELASTICITY E Dim: $L^{-1}MT^{-2}$
module de l'élasticité longitudinale; (*module de Young*)
Elastizitätsmodul
SI unit: pascal Pa
Note: also called ('Young modulus'); in BRD N/m^2 is used

MODULUS OF IMPEDANCE $|Z|$ Dim: $L^{2}MT^{-3}I^{-2}$
module de l'impédance, (*impédance*)
Scheinwiderstand, Betrag der Impedanz
SI unit: ohm Ω
M/S: MΩ, kΩ, mΩ
Note: also called ('impedance')

MODULUS OF RIGIDITY see: shear modulus

MODULUS OF SECTION see: section modulus

MOLALITY OF SOLUTE SUBSTANCE B b_B, m_B Dim: $M^{-1}N$
molalité du soluté B
Molalität eines gelösten Stoffes B
SI unit: mole per kilogram mol/kg
M/S: mmol/kg

MOLAR ABSORPTION COEFFICIENT κ Dim: $L^{2}N^{-1}$
coefficient d'absorption molaire
stoffmengenbezogener (*molarer*) *Absorptionskoeffizient*
SI unit: square metre per mole m^2/mol

MOLAR CONDUCTIVITY Λ_m Dim: $M^{-1}T^3I^2N^{-1}$
conductivité molaire
konzentrationsbezogene (molare) Leitfähigkeit
SI unit: siemens square metre per mole S·m²/mol

MOLAR ENTROPY S_m Dim: $L^2MT^{-2}\Theta^{-1}N^{-1}$
entropie molaire
stoffmengenbezogene (molare) Entropie
SI unit: joule per mole kelvin J/(mol·K)

MOLAR FLOW RATE q_v Dim: $T^{-1}N$
débit molaire
Stoffmengenstrom, Stoffmengendurchfluß, stoffmengenbezogener (molarer) Durchfluß
SI unit: mole per second mol/s

MOLAR GAS CONSTANT R Dim: $L^2MT^{-2}\Theta^{-1}N^{-1}$
constante molaire des gaz
universelle Gaskonstante, molare Gaskonstante
SI unit: joule per mole kelvin J/(mol·K)
$R = 8.314\,41$ J/(mol·K)

MOLAR HEAT CAPACITY C_m Dim: $L^2MT^{-2}\Theta^{-1}N^{-1}$
capacité thermique molaire
stoffmengenbezogene (molare) Wärmekapazität
SI unit: joule per mole kelvin J/(mol·K)

MOLAR INTERNAL ENERGY U_m, (E_m) Dim: $L^2MT^{-2}N^{-1}$
énergie interne molaire
stoffmengenbezogene (molare) innere Energie
SI unit: joule per mole J/mol
M/S: kJ/mol

MOLAR MASS M Dim: MN^{-1}
masse molaire
stoffmengenbezogene (molare) Masse
SI unit: kilogram per mole kg/mol
M/S: g/mol

MOLAR VOLUME V_m Dim: L^3N^{-1}
volume molaire
stoffmengenbezogenes (molares) Volumen
SI unit: cubic metre per mole m^3/mol
M/S: dm^3/mol, cm^3/mol
oSI unit: litre per mole l/mol, L/mol

MOLARITY
molarité
Molarität
Note: obsol. – see: 'concentration of substance B'; not to be confused with 'molality'

MOLE FRACTION OF SUBSTANCE B x_B, (y_B) Dim: 1
fraction molaire du constituant B
Stoffmengenanteil (Stoffmengengehalt) eines Stoffes B; (Stoffmengenbruch; Molenbruch)
Note: also called 'amount-of-substance fraction'; 'mole fraction in %' is to replace 'mole %'

MOLE PERCENT (mole %)
pour cent molaire (% mol)
Molprozent (Mol. %)
Note: see 'mole fraction of substance B'

MOLE RATIO OF SOLUTE SUBSTANCE B r_B Dim: 1
rapport molaire du soluté B
Stoffmengenverhältnis eines gelösten Stoffes B
Note: also called 'amount-of-substance ratio'

MOLECULAR CONCENTRATION OF SUBSTANCE B C_B Dim: L^{-3}
concentration moléculaire du constituant B
Molekülkonzentration eines Stoffes B
SI unit: reciprocal cubic metre m^{-3}

MOLECULAR FLUX see: molecular flow rate

MOLECULAR PARTITION FUNCTION q Dim: 1
fonction de partition moléculaire
molekulare Zustandssume, Molekülzustandssumme
Note: also called 'partition function of a molecule'

MOLECULAR WEIGHT
poids moléculaire
Molekulargewicht
Note: obsol. – see: 'relative molecular mass'

MOLECULE CONDUCTANCE C_N, U_N Dim: L^3T^{-1}
conductance-molécules
Teilchenströmungsleitwert
SI unit: cubic metre per second m^3/s
oSI unit: litre per second l/s, L/s
Note: the indicated German term is used in preference to *Molekülströmungsleitwert*

MOLECULE FLOW RATE q_N Dim: T^{-1}
débit-molécules, flux de molécules
Teilchendurchfluß; Teilchenstrom
SI unit: reciprocal second s^{-1}
Note: also called 'molecular flux'; the indicated German terms are used in preference to *Moleküldurchfluß*

MOLECULE FLOW RATE DENSITY (—) Dim: $L^{-2}T^{-1}$
débit-molécules surfacique, densité de flux de molécules
Teilchendurchflußdichte; Teilchenstromdichte
SI unit: reciprocal second reciprocal square metre
$s^{-1} \cdot m^{-2}$
Note: also called 'density of molecular flux'; the indicated German terms are used in preference to *Moleküldurchflußdichte*

MOMENT see: electric dipole m., electromagnetic m., magnetic dipole m., magnetic m. . . ., nuclear quadrupole m., second m. . . ., second polar m. . . .

MOMENT OF A COUPLE see: torque

MOMENT OF FORCE M Dim: L^2MT^{-2}
moment d'une force
Moment einer Kraft, Kraftmoment
SI unit: newton metre N·m
M/S: MN·m, kN·m, mN·m, μN·m
Note: The rider 'at a point' (*en un point*; *in einem Punkt*) is sometimes added

MOMENT OF INERTIA I, J Dim: L^2M
moment d'inertie; (*moment d'inertie dynamique*)
Trägheitsmoment, Massenmoment 2. Grades; (formerly: *Massenträgheitsmoment*)
SI unit: kilogram metre squared kg·m²
Note: also called ('dynamic moment of inertia'); the rider 'about an axis' (*par rapport à un axe*; *um eine Achse*) is sometimes used

MOMENT OF MOMENTUM L Dim: L^2MT^{-1}
moment cinétique, moment de quantité de mouvement
Drall, Drehimpuls
SI unit: kilogram metre squared per second kg·m²/s
Note: also called 'angular momentum'

MOMENTUM p Dim: LMT^{-1}
quantité de mouvement
Bewegungsgröße, Impuls; (Kraftstoß)
SI unit: kilogram metre per second kg·m/s
Note: also called 'linear momentum' to distinguish it from 'angular momentum' (q.v.)

MULTIPLICATION FACTOR k Dim: 1
facteur de multiplication
Multiplikationsfaktor, Vermehrungsfaktor

MUTUAL INDUCTANCE M, L_{12} Dim: $L^2MT^{-2}I^{-2}$
inductance mutuelle
gegenseitige Induktivität
SI unit: henry H
M/S: mH, μH, nH, pH

N

NÉEL TEMPERATURE T_N Dim: Θ
température de Néel
Néel-Temperatur
SI unit: kelvin K

NEUTRON EXCESS NUMBER $(N–Z)$ Dim: 1
excès de neutrons
Neutronenüberschuß
Note: also called 'neutron excess' or ('isotopic number')

NEUTRON FLUENCE RATE ϕ Dim: $L^{-2}T^{-1}$
débit de fluence de neutrons
Neutronenflußdichte
SI unit: reciprocal second reciprocal square metre
$s^{-1} \cdot m^{-2}$
Note: also called 'neutron flux density' or ('neutron flux')

NEUTRON NUMBER N Dim: 1
nombre de neutrons
Neutronenzahl

NEUTRON NUMBER DENSITY n Dim: L^{-3}
nombre volumique de neutrons
Neutronen(an)zahldichte
SI unit: reciprocal cubic metre m^{-3}

NEUTRON SPEED v Dim: LT^{-1}
vitesse de neutrons
Neutronengeschwindigkeit
SI unit: metre per second m/s

NEUTRON YIELD PER ABSORPTION η Dim: 1
nombre des neutrons produits par neutron absorbé; (*facteur* η, *facteur êta*)
Neutronenausbeute je Absorption
Note: also called (η-factor)

NEUTRON YIELD PER FISSION ν Dim: 1
nombre des neutrons produits par fission; (*facteur* ν, factuer nu)
Neutronenausbeute je Spaltung
Note: also called (ν-factor)

NON-LEAKAGE PROBABILITY Λ Dim: 1
probabilité de non-fuite
Verbleibwahrscheinlichkeit

NORMAL STRESS σ Dim: $L^{-1}MT^{-2}$

contrainte normale, tension normale
Normalspannung
SI unit: pascal Pa
M/S: GPa, MPa, kPa; N/mm^2
Note: in BRD N/m^2 is used

NUCLEAR MAGNETON μ_N Dim: L^2I

magnéton nucléaire
Kernmagneton
SI unit: ampere square metre $A \cdot m^2$
$\mu_N = 5.050\,824 \times 10^{-27}\ A \cdot m^2$
Note: $A \cdot m^2 = J/T$

NUCLEAR PRECESSION ANGULAR FREQUENCY ω_N Dim: T^{-1}

pulsation de précession nucléaire de Larmor
Kreisfrequenz der Kernpräzession
SI unit: reciprocal second s^{-1}
oSI unit: radian per second rad/s
Note: also called 'nuclear precession circular frequency'

NUCLEAR QUADRUPOLE MOMENT Q Dim: L^2

moment quadripolaire nucléaire
Kernquadrupolmoment
SI unit: square metre m^2

NUCLEAR RADIUS R Dim: L

rayon nucléaire, rayon du noyau
Kernradius
SI unit: metre m

NUCLEAR SPIN QUANTUM NUMBER I Dim: 1

nombre quantique de spin nucléaire
Kernspin-Quantenzahl

NUCLEON NUMBER A Dim: 1

nombre de nucléons, nombre de masse
Nukleonenzahl, Massenzahl
Note: also called 'mass number'; this quantity is defined as the number of nucleons in an atomic nucleus and should therefore be called 'nucleon number'

NUCLIDIC MASS see: mass of atom . . .

NU-FACTOR see: neutron yield per fission

NUMBER see: Alfvén n., atomic n., Biot n., charge n. . . ., Cowling n., Euler n., Fourier n., Froude n., Grashof n., Hartmann n., hyperfine structure quantum n., Knudsen n., Lewis n., Mach n., magnetic quantum n., magnetic Reynolds n., Margoulis n., neutron excess n., neutron n., nuclear spin quantum n., nucleon n., Nusselt n., orbital angular momentum quantum n., Péclet n., Prandtl n., principal quantum n., proton n., Rayleigh n., Reech n., Reynolds n., Schmidt n., Sherwood n., spin angular momentum quantum n., Stanton n., stoichiometric n. . .., Strouhal n., total angular momentum quantum n., Weber n.

NUMBER DENSITY OF MOLECULES (OR PARTICLES) n Dim: L^{-3}
nombre volumique de molécules (ou de particules)
volumenbezogene Molekül(an)zahl; volumenbezogene Teilchen(an)zahl; Molekül(an)zahldichte; Teilchen(an) zahldichte
SI unit: reciprocal cubic metre m^{-3}

NUMBER OF MOLECULES (OR PARTICLES) N Dim: 1
nombre de molécules (ou de particules)
Anzahl der Moleküle (oder Teilchen); Molekül(an)zahl; (Teilchen(an)zahl)

NUMBER OF PAIRS OF POLES p Dim: 1
nombre de paires de pôles
Anzahl der Polpaare

NUMBER OF PHASES m Dim: 1
nombre de phases
Anzahl der Phasen

NUMBER OF TURNS IN A WINDING N Dim: 1
nombre de tours (spires) de l'enroulement
Windungszahl

NUSSELT NUMBER Nu Dim: 1
nombre de Nusselt
Nußelt-Zahl

NUSSELT NUMBER FOR MASS TRANSFER Nu^* Dim: 1
nombre de Nusselt pour transfert de masse
Nußelt-Zahl der Stoffübertragung
Note: also called 'Sherwood number'

O

ORBITAL ANGULAR MOMENTUM QUANTUM NUMBER l_i, L Dim: 1
nombre quantique du moment cinétique orbital
Bahndrehimpuls-Quantenzahl

ORDER OF REFLEXION n Dim: 1
ordre de réflexion
Reflexionsordnung

OSMOTIC COEFFICIENT OF SOLVENT SUBSTANCE A ϕ Dim: 1
coefficient osmotique du solvant A; facteur osmotique du solvant A
osmotischer Koeffizient eines Lösungsmittels A
Note: particularly in a dilute liquid solution; this quantity is a 'factor' rather than a 'coefficient'

OSMOTIC PRESSURE Π Dim: $L^{-1}MT^{-2}$
pression osmotique
osmotischer Druck
SI unit: pascal Pa

P

PACKING FRACTION f Dim: 1
facteur de tassement
Packungsanteil

PARAMETER see: Grüneisen p., Landau-Ginzburg p., long range order p., short range order p.

PARTIAL PRESSURE OF SUBSTANCE B Dim: $L^{-1}MT^{-2}$
(in a gaseous mixture) p_B
pression partielle du constituant B (dans un mélange gazeux)
Partialdruck eines Stoffes B (in einem Gasgemisch)
SI unit: pascal Pa

PARTICLE . . . see also: sound particle

PARTICLE FLUENCE Φ Dim: L^{-2}
fluence de particules
Teilchenfluenz, Fluenz
SI unit: reciprocal square metre m^{-2}

PARTICLE FLUENCE RATE ϕ Dim: $L^{-2}T^{-1}$
débit de fluence de particules
Teilchenflußdichte, Flußdichte
SI unit: reciprocal square metre reciprocal second $m^{-2} \cdot s^{-1}$
Note: also called 'particle flux density'

PÉCLET NUMBER *Pe* Dim: 1
nombre de Péclet
Péclet-Zahl

PÉCLET NUMBER FOR MASS TRANSFER *Pe** Dim: 1
nombre de Péclet pour transfert de masse
Péclet-Zahl der Stoffübertragung
Note: also called 'Bodenstein number'

PELTIER COEFFICIENT FOR SUBSTANCES a AND b Π_{ab} Dim: $L^2MT^{-3}I^{-1}$
coefficient de Peltier pour deux substances a et b
Peltier-Koeffizient für Stoffe a und b
SI unit: volt V

PERCENT see: mass p., mole p., volume p., weight p.

PERIOD T Dim: T
période
Periodendauer
SI unit: second s
M/S: ms, μs
Note: also called 'periodic time'; cf. 'time'

PERMEABILITY μ Dim: $LMT^{-2}I^{-2}$
perméabilité; perméabilité absolue–IEC
Permeabilität
SI unit: henry per metre H/m
M/S: μH/m, nH/m
Note: also called 'absolute permeability' – IEC

PERMEABILITY OF VACUUM μ_0 Dim: $LMT^{-2}I^{-2}$
perméabilité du vide, constante magnétique
magnetische Feldkonstante; (Permeabilität des leeren Raumes)
SI unit: henry per metre H/m
$\mu_0 = 1.256\,637\,061\,44 \times 10^{-6}$ H/m
Note: also called 'magnetic constant' which may have the symbol Γ_m if it is to be distinguished from permeability of vacuum

PERMEANCE Λ, (P) Dim: $L^2MT^{-2}I^{-2}$
perméance
magnetischer Leitwert, Permeanz
SI unit: henry H
Note: the reciprocal of permeance is called 'reluctance'

PERMITTIVITY ε Dim: $L^{-3}M^{-1}T^4I^2$
permittivité; permittivité absolue – IEC
Permittivität; (Dielektrizitätskonstante)
SI unit: farad per metre F/m
M/S: μF/m, nF/m, pF/m
Note: also called 'absolute permittivity' – IEC, or ('capacitivity' – IEC)

PERMITTIVITY OF VACUUM ε_0 Dim: $L^{-3}M^{-1}T^4I^2$

permittivité du vide, constante électrique
elektrische Feldkonstante, (*Dielektrizitätskonstante des leeren Raumes*)
SI unit: farad per metre F/m
$\varepsilon_0 = 8.854\,187\,82 \times 10^{-12}$ F/m
Note: also called 'electric constant'

PHASE COEFFICIENT β Dim: L^{-1}

déphasage linéique; (*constante de phase*)
Phasenkoeffizient, Phasenbelag; (*Phasenkonstante*)
SI unit: reciprocal metre m^{-1}
Note: also called ('phase-change coefficient') and ('phase constant')

PHASE DIFFERENCE ϕ, (θ) Dim: 1

déphasage, différence de phase
Phasenverschiebungswinkel, Phasenwinkel
SI unit: radian rad
Note: also called 'phase displacement'

PLANCK CONSTANT h Dim: L^2MT^{-1}

constante de Planck
Planck-Konstante; Plancksches Wirkungsquantum
SI unit: joule second J·s
$h = 6.626\,176 \times 10^{-34}$ J·s
$\hbar = h/2\pi = 1.054\,588\,7 \times 10^{-34}$ J·s
Note: J·s = J/Hz; $\hbar$ is sometimes called 'Dirac constant' q.v.

PLANCK FUNCTION Y Dim: $L^2MT^{-2}\Theta^{-1}$

fonction de Planck
Planck-Funktion
SI unit: joule per kelvin J/K

POISSON RATIO μ, ν Dim: 1

coefficient de Poisson, nombre de Poisson
Poisson-Zahl
Note: also called 'Poisson number'

POTENTIAL see: chemical p., electric p., magnetic vector p.

POTENTIAL DIFFERENCE U, (V) Dim: $L^2MT^{-3}I^{-1}$
différence de potentiel, tension
elektrische Potentialdifferenz, elektrische Spannung
SI unit: volt V
M/S: MV, kV, mV, μV
Note: also called 'tension' or 'voltage' (IEC)

POTENTIAL ENERGY E_p, V, Φ Dim: L^2MT^{-2}
énergie potentielle
potentielle Energie
SI unit: joule J
Note: cf. 'energy'

POWER see: active p., apparent p., radiant p., reactive p., relative linear stopping p., relative mass stopping p., total atomic stopping p., total linear stopping p., total mass stopping p.

POWER P Dim: L^2MT^{-3}
puissance
Leistung
SI unit: watt W
M/S: TW, GW, MW, kW, mW, μW, nW

POWER FACTOR λ Dim: 1
facteur de puissance
Leistungsfaktor

POWER OF A LENS D Dim: L^{-1}
vergence
Brechwert; Brechkraft
SI unit: reciprocal metre m^{-1}
other unit: dioptre δ, dpt

POYNTING VECTOR S Dim: MT^{-3}
vecteur de Poynting
Poynting-Vektor
SI unit: watt per square metre W/m^2

PRANDTL NUMBER *Pr* Dim: 1
nombre de Prandtl
Prandtl-Zahl

PRESSURE see: absolute p., ambient p., differential p., gauge p., osmotic p., partial p., saturation vapour p., sound p., standard p., static p., total p.

PRESSURE *p* Dim: $L^{-1}MT^{-2}$
pression
Druck
SI unit: pascal Pa
M/S: GPa, MPa, kPa, mPa, μPa
oSI unit (for pressure of fluid): bar bar (mbar, μbar)

PRESSURE COEFFICIENT *β* Dim: $L^{-1}MT^{-2}\Theta^{-1}$
coefficient de pression
Druckkoeffizient
SI unit: pascal per kelvin Pa/K

PRINCIPAL QUANTUM NUMBER *n* Dim: 1
nombre quantique principal
Haupt-Quantenzahl

PROBABILITY see: non-leakage p., resonance escape p., sticking p.

PROPAGATION COEFFICIENT *γ* Dim: L^{-1}
exposant linéique de propagation; (*constante de propagation*)
Ausbreitungskoeffizient; (*Fortpflanzungskonstante*)
SI unit: reciprocal metre m^{-1}
Note: also called ('propagation constant')

PROTON NUMBER *Z* Dim: 1
nombre de protons
Protonenzahl, Kernladungszahl, Ordnungszahl
Note: also called 'atomic number'; this quantity is defined as the number of protons in an atomic nucleus and should therefore be called 'proton number'

PULSATANCE see: circular frequency

Q

QUANTITY OF GAS (pressure-volume units) G Dim: L^2MT^{-2}
quantité énergétique de gaz
pV-Wert
SI unit: pascal cubic metre Pa·m³

QUANTITY OF ELECTRICITY see: electric charge

QUANTITY OF HEAT see: heat

QUANTITY OF ILLUMINATION see: light exposure

QUANTITY OF LIGHT Q, (Q_v) Dim: TJ
quantité de lumière
Lichtmenge
SI unit: lumen second lm·s
oSI unit: lumen hour lm·h

R

RADIANCE L, (L_e) Dim: MT^{-3}
luminance énergétique, radiance
Strahldichte
SI unit: watt per steradian square metre W/(sr·m²)

RADIANT EMITTANCE see: radiant exitance

RADIANT ENERGY Q, W, (U, Q_e) Dim: L^2MT^{-2}
énergie rayonnante
Strahlungsenergie, Strahlungsmenge
SI unit: joule J

RADIANT ENERGY DENSITY w, (u) Dim: $L^{-1}MT^{-2}$
énergie rayonnante volumique
Strahlungsenergiedichte
SI unit: joule per cubic metre J/m³

RADIANT ENERGY FLUENCE RATE ϕ, ψ Dim: MT^{-3}
débit de fluence énergétique; densité de flux énergétique
Strahlungsflußdichte
SI unit: watt per square metre W/m^2
Note: also called ('radiant flux density')

RADIANT ENERGY FLUX see: radiant power

RADIANT EXITANCE M, (M_e) Dim: MT^{-3}
exitance énergétique; (formerly: *émittance énergétique)*
spezifische Ausstrahlung
SI unit: watt per square metre W/m^2
Note: formerly called ('radiant emittance')

RADIANT EXPOSURE (H_e, H) Dim: MT^{-2}
exposition énergétique
Bestrahlung
SI unit: joule per square metre J/m^2

RADIANT FLUX DENSITY see: radiant energy fluence rate

RADIANT INTENSITY I, (I_e) Dim: L^2MT^{-3}
intensité énergétique
Strahlstärke
SI unit: watt per steradian W/sr

RADIANT POWER P, Φ, (Φ_e) Dim: L^2MT^{-3}
puissance rayonnante, flux énergétique
Strahlungsfluß, Strahlungsleistung
SI unit: watt W
Note: also called 'radiant energy flux'

RADIUS see: Bohr r., electron r., nuclear r.

RADIUS r Dim: L
rayon
Halbmesser, Radius
SI unit: metre m
M/S: km, cm, mm, µm, nm, pm, fm

RANGE see: mass linear r., mean mass r.

RATE see: absorbed dose r., collision r., diffusion coefficient for neutron fluence r., energy fluence r., exposure r., heat flow r., impingement r., ion dose r., kerma r., leak r., mass flow r., molar flow r., molecule flow r., neutron fluence r., particle fluence r., radiant energy fluence r., volume collision r., volume flow r.

RATIO see: mobility r., mole r. . . ., Poisson r., thermal diffusion r., turns r.

RATIO OF SPECIFIC HEAT CAPACITIES γ Dim: 1
rapport des capacités thermiques massiques
Verhältnis der spezifischen Wärmekapazitäten

RAYLEIGH NUMBER Ra Dim: 1
nombre de Rayleigh
Rayleigh-Zahl

REACTANCE X Dim: $L^2MT^{-3}I^{-2}$
réactance
Blindwiderstand, Reaktanz
SI unit: ohm Ω
M/S: MΩ, kΩ, mΩ

REACTION ENERGY Q Dim: L^2MT^{-2}
énergie de réaction
Reaktionsenergie
SI unit: joule J
oSI unit: electronvolt eV

REACTIVE POWER Q, (P_q) Dim: L^2MT^{-3}
puissance réactive
Blindleistung
SI unit: watt W
other unit: var var

REACTIVITY ρ Dim: 1
réactivité
Reaktivität

REACTOR TIME CONSTANT T Dim: T
constante de temps du réacteur; (période du réacteur)
Reaktorzeitkonstante; (Reaktorperiode)
SI unit: second s
Note: also called ('reactor period')

RECOMBINATION COEFFICIENT α Dim: L^3T^{-1}
coefficient de recombinaison
Rekombinationskoeffizient
SI unit: cubic metre per second m^3/s

REECH NUMBER Dim: 1
nombre de Reech
Reech-Zahl
Note: alternative name for 'Froude number'

REFLECTION COEFFICIENT (of sound) r, ρ Dim: 1
facteur de réflexion
Schallreflexionsgrad

REFRACTIVE INDEX n Dim: 1
indice de réfraction
Brechzahl; (Brechungszahl)

RELATIVE ACTIVITY . . . see: activity . . .

RELATIVE ATOMIC MASS (OF AN ELEMENT) A_r Dim: 1
masse atomique relative (d'un élément)
relative Atommasse (eines Elementes)
Note: formerly called ('atomic weight')

RELATIVE CAPACITIVITY see: relative permittivity

RELATIVE DENSITY d Dim: 1
densité relative; densité
relative Dichte
Note: formerly called 'specific gravity'

RELATIVE ELONGATION see: linear strain

RELATIVE LINEAR STOPPING POWER ($S_{l,\mathrm{r}}$) Dim: 1
pouvoir d'arrêt linéique relatif
relatives lineares Bremsvermögen

RELATIVE MASS DEFECT B_r Dim: 1
défaut de masse relatif
relativer Massendefekt

RELATIVE MASS EXCESS Δ_r Dim: 1
excès de masse relatif
relativer Massenüberschuß

RELATIVE MASS STOPPING POWER ($S_{m,\mathrm{r}}$) Dim: 1
pouvoir d'arrêt massique relatif
relatives Massenbremsvermögen

RELATIVE MOLECULAR MASS (OF A SUBSTANCE) M_r Dim: 1
masse moléculaire relative (d'un corps)
relative Molekülmasse (eines Stoffes); relative Molekularmasse
Note: formerly called ('molecular weight')

RELATIVE PERMEABILITY μ_r Dim: 1
perméabilité relative; facteur de perméabilité – IEC
Permeabilitätszahl

RELATIVE PERMITTIVITY ε_r Dim: 1
permittivité relative; facteur de permittivité – IEC
Permittivitätszahl; (Dielektrizitätszahl)
Note: also called ('relative capacitivity') – IEC

RELATIVE PRESSURE COEFFICIENT α_p Dim: Θ^{-1}
coefficient relatif de pression
(thermischer) Spannungskoeffizient
SI unit: reciprocal kelvin K^{-1}

RELAXATION TIME τ Dim: T
temps de relaxation
Relaxationszeit
SI unit: second s

RELUCTANCE R, R_m Dim: $L^{-2}M^{-1}T^2I^2$
réluctance
magnetischer Widerstand, Reluktanz
SI unit: reciprocal henry H^{-1}
Note: the reciprocal of reluctance is called 'permeance'

REPETENCY see: wavenumber

RESIDENCE TIME τ Dim: T
temps de séjour
Verweildauer
SI unit: second s

RESISTANCE (electrical) R Dim: $L^2MT^{-3}I^{-2}$
résistance
elektrischer Widerstand, Resistanz
SI unit: ohm Ω
M/S: GΩ, MΩ, kΩ, mΩ, μΩ
Note: the reciprocal of resistance is called 'conductance'; both resistance to direct current and resistance as the real part of impedance have the same symbol, dimension and unit

RESISTANCE (fluid flow) w Dim: $L^{-3}T$
résistance
Strömungswiderstand
SI unit: second per cubic metre s/m^3
oSI unit: second per litre s/l, s/L
Note: the reciprocal of 'conductance'

RESISTIVITY ρ Dim: $L^3MT^{-3}I^{-2}$
résistivité
spezifischer elektrischer Widerstand, Resistivität
SI unit: ohm metre Ω·m
M/S: GΩ·m, MΩ·m, kΩ·m, Ω·cm, mΩ·m, μΩ·m, nΩ·m
Note: the reciprocal of resistivity is called 'conductivity'. Formerly called 'specific resistance'

RESONANCE ENERGY E_r, E_{res} Dim: L^2MT^{-2}
énergie de résonance
Resonanzenergie
SI unit: joule J
oSI unit: electronvolt eV

RESONANCE ESCAPE PROBABILITY p Dim: 1
facteur antitrappe; (formerly: *probabilité d'échappement de résonance*)
Bremsnutzung; (*Resonanz-Entkommwahrscheinlichkeit*)

(REST) MASS OF ELECTRON m_e Dim: M
masse (au repos) de l'électron
Elektronen(ruh)masse, Ruhmasse des Elektrons
SI unit: kilogram kg
oSI unit: (unified) atomic mass unit u
$m_e = 9.109\,534 \times 10^{-31}$ kg
$= 5.485\,802\,6 \times 10^{-4}$ u

(REST) MASS OF NEUTRON m_n Dim: M
masse (au repos) du neútron
Neutronen(ruh)masse, Ruhmasse des Neutrons
SI unit: kilogram kg
oSI unit: (unified) atomic mass unit u
$m_n = 1.674\,954\,3 \times 10^{-27}$ kg
$= 1.008\,665\,012$ u

(REST) MASS OF PROTON m_p Dim: M
masse (au repos) du proton
Protonen(ruh)masse. Ruhmasse des Protons
SI unit: kilogram kg
oSI unit: (unified) atomic mass unit u
$m_p = 1.672\,648\,5 \times 10^{-27}$ kg
$= 1.007\,276\,470$ u

REVERBERATION TIME T Dim: T
durée de réverbération
Nachhallzeit
SI unit: second s

REYNOLDS NUMBER *Re* Dim: 1

nombre de Reynolds
Reynolds-Zahl

RICHARDSON CONSTANT *A* Dim: $L^{-2}I\Theta^{-2}$

facteur de Richardson
Richardson-Konstante
SI unit: ampere per square metre kelvin squared
$A/(m^2 \cdot K^2)$

ROTATIONAL FREQUENCY *n* Dim: T^{-1}

fréquence de rotation
Drehzahl, Umdrehungsfrequenz
SI unit: reciprocal second s^{-1}
other units: revolutions per minute r/min,
revolutions per second r/s

RYDBERG CONSTANT R_∞ Dim: L^{-1}

constante de Rydberg
Rydberg-Konstante
SI unit: reciprocal metre m^{-1}
$R_\infty = 1.097\,373\,177 \times 10^7\ m^{-1}$

RYDBERG ENERGY (Ry) L^2MT^{-2}

énergie de Rydberg
Rydberg-Energie
$Ry = R_\infty \cdot h \cdot c = 2.179\,907\,2 \times 10^{-18}\ J$
$= 1.360\,580\,4 \times 10\ eV$

S

SATURATION VAPOUR PRESSURE p_L Dim: $L^{-1}MT^{-2}$

pression de vapeur saturante, pression de saturation
Sättigungsdampfdruck
SI unit: pascal Pa

SCATTERING CROSS SECTION σ_s, σ_S Dim: L^2
section efficace de diffusion
Streuquerschnitt
SI unit: square metre m^2
other unit: barn b

SCHMIDT NUMBER *Sc* Dim: 1
nombre de Schmidt
Schmidt-Zahl

SECOND (AXIAL) MOMENT OF AREA I_a, (I) Dim: L^4
moment quadratique (axial) d'une aire plane
(axiales) Flächenmoment 2. Grades; (formerly: *axiales Flächenträgheitsmoment*)
SI unit: metre to the fourth power m^4

SECOND POLAR MOMENT OF AREA I_p Dim: L^4
moment quadratique polaire d'une aire plane
polares Flächenmoment 2. Grades; (formerly: *polares Flächenträgheitsmoment*)
SI unit: metre to the fourth power m^4

SECOND RADIATION CONSTANT c_2 Dim: $L\Theta$
seconde constante de rayonnement
zweite (Plancksche) Strahlungskonstante
SI unit: metre kelvin m·K
$c_2 = 1.438\,786 \times 10^{-2}$ m·K
Note: The unit kelvin-mètre K·m is sometimes used in France to avoid possible confusion with millikelvin mK

SECTION MODULUS *Z, W* Dim: L^3
module d'inertie
Widerstandsmoment
SI unit: metre cubed m^3
Note: also called ('modulus of section')

SEEBECK COEFFICIENT FOR SUBSTANCES a AND b S_{ab}, ε_{ab} Dim: $L^2MT^{-3}I^{-1}\Theta^{-1}$
coefficient de Seebeck pour deux substances a et b
Seebeck-Koeffizient für Stoffe a und b
SI unit: volt per kelvin V/K

SELF INDUCTANCE L Dim: $L^2MT^{-2}I^{-2}$
inductance propre
Induktivität, Selbstinduktivität
SI unit: henry H
M/S: mH, μH, nH, pH

SHEAR MODULUS G Dim: $L^{-1}MT^{-2}$
module d'élasticité de glissement; (*module de Coulomb*)
Schubmodul
SI unit: pascal Pa
Note: also called 'modulus of rigidity' and ('Coulomb modulus'); in BRD N/m² is used

SHEAR STRAIN γ Dim: 1
glissement unitaire
Schiebung
Note: in BRD radian is used as a unit of shear strain

SHEAR STRESS τ Dim: $L^{-1}MT^{-2}$
contrainte tangentielle; (*tension de cisaillement*)
Schubspannung
SI unit: pascal Pa
Note: in BRD N/m² is used

SHERWOOD NUMBER Sh Dim: 1
nombre de Sherwood
Sherwood-Zahl
Note: alternative name for 'Nusselt number for mass transfer'

SHORT RANGE ORDER PARAMETER σ Dim: 1
paramètre d'ordre local
Nahordnungsparameter

SLOWING-DOWN AREA L_s^2, L_{sl}^2 Dim: L^2
aire de ralentissement
Bremsfläche
SI unit: metre squared m²

SLOWING-DOWN DENSITY q Dim: $L^{-3}T^{-1}$
densité de ralentissement
Bremsdichte
SI unit: reciprocal second reciprocal cubic metre
$s^{-1} \cdot m^{-3}$

SLOWING-DOWN LENGTH L_s, L_{sl} Dim: L
longueur de ralentissement
Bremslänge
SI unit: metre m

SOLAR CONSTANT (—) Dim: MT^{-3}
constante solaire
Solarkonstante
1.39×10^3 J/(s·m²) or W/m²

SOLID ANGLE Ω Dim: 1
angle solide
Raumwinkel, (räumlicher Winkel)
SI unit: steradian sr

SOUND ... see also: acoustic ...
Note: many quantities of acoustics are used without the qualifier 'sound'. See, e.g., attenuation coeff., dissipation coeff., phase coeff., propagation coeff., reflection coeff., transmission coeff.

SOUND ENERGY DENSITY w, (w_a), (E) Dim: $L^{-1}MT^{-2}$
énergie volumique acoustique
Schallenergiedichte
SI unit: joule per cubic metre J/m³

SOUND ENERGY FLUX P, (P_a) Dim: L^2MT^{-3}
flux d'énergie acoustique, puissance acoustique
Schalleistung, (Schallenergiefluß)
SI unit: watt W
M/S: kW, mW, μW, pW
Note: also called 'sound power'

SOUND INTENSITY I, J Dim: MT^{-3}
intensité acoustique
Schallintensität
SI unit: watt per square metre W/m²
M/S: mW/m², μW/m², pW/m²

SOUND INTENSITY LEVEL L_I Dim: 1
niveau d'intensité acoustique
Schallintensitätspegel
SI unit: —
other unit: decibel dB

SOUND PARTICLE ACCELERATION a Dim: LT^{-2}
accélération acoustique d'une particule
Teilchenbeschleunigung
SI unit: metre per second squared m/s^2
Note: often qualified by terms such as instantaneous, maximum, or root mean square

SOUND PARTICLE DISPLACEMENT ξ, (x) Dim: L
élongation d'une particule
Schallausschlag; Teilchenausschlag
SI unit: metre m
Note: See note to 'sound particle acceleration'

SOUND PARTICLE VELOCITY u, v Dim: LT^{-1}
vitesse acoustique d'une particule
Schallschnelle
SI unit: metre per second m/s
M/S: mm/s
Note: see note to 'sound particle acceleration'; cf. 'velocity of sound'

SOUND POWER see: sound energy flux

SOUND POWER LEVEL L_P, L_W Dim: 1
niveau de puissance acoustique
Schalleistungspegel
SI unit: —
other unit: decibel dB

SOUND PRESSURE p, (p_a) Dim: $L^{-1}MT^{-2}$

pression acoustique
Schalldruck
SI unit: pascal Pa
M/S: mPa, μPa
other unit: bar bar
Note: see note to 'sound particle acceleration'

SOUND PRESSURE LEVEL L_p Dim: 1

niveau de pression acoustique
Schalldruckpegel
SI unit: —
other unit: decibel dB

SOUND REDUCTION INDEX R Dim: 1

indice d'affaiblissement acoustique
Schalldämm-Maß
SI unit: —
other unit: decibel dB
Note: also called 'sound transmission loss'

SOUND TRANSMISSION LOSS see: sound reduction index

SPECIFIC ACOUSTIC IMPEDANCE Z_s Dim: $L^{-2}MT^{-1}$

impédance acoustique spécifique
spezifische Schallimpedanz, Feldimpedanz
SI unit: pascal second per metre Pa·s/m

SPECIFIC ACTIVITY a Dim: $M^{-1}T^{-1}$

activité massique
spezifische Aktivität
SI unit: becquerel per kilogram Bq/kg
other unit: curie per kilogram Ci/kg
Note: specific activity of a radioactive substance

SPECIFIC ENTHALPY h, (i) Dim: L^2T^{-2}

enthalpie massique
spezifische Enthalpie
SI unit: joule per kilogram J/kg

SPECIFIC ENTROPY s Dim: $L^2T^{-2}\Theta^{-1}$
entropie massique
spezifische Entropie
SI unit: joule per kilogram kelvin J/(kg·K)
M/S: kJ/(kg·K)

SPECIFIC GAMMA RAY CONSTANT Γ Dim: $L^2M^{-1}TI$
constante spécifique de rayonnement gamma
spezifische Gammastrahlenkonstante
SI unit: coulomb metre squared per kilogram
C·m²/kg
other unit: röntgen metre squared per curie hour
R·m²/(Ci·h)

SPECIFIC GIBBS FREE ENERGY g Dim: L^2T^{-2}
enthalpie libre massique
spezifische freie Enthalpie
SI unit: joule per kilogram J/kg
Note: also called 'specific Gibbs function'

SPECIFIC GRAVITY see: relative density

SPECIFIC HEAT CAPACITY c Dim: $L^2T^{-2}\Theta^{-1}$
capacité thermique massique
spezifische Wärmekapazität
SI unit: joule per kilogram kelvin J/(kg·K)
M/S: kJ/(kg·K)

SPECIFIC HEAT CAPACITY AT CONSTANT PRESSURE c_p Dim: $L^2T^{-2}\Theta^{-1}$
capacité thermique massique à pression constante
spezifische Wärmekapazität bei konstantem Druck
SI unit: joule per kilogram kelvin J/(kg·K)

SPECIFIC HEAT CAPACITY AT CONSTANT VOLUME c_V Dim: $L^2T^{-2}\Theta^{-1}$
capacité thermique massique à volume constant
spezifische Wärmekapazität bei konstantem Volumen
SI unit: joule per kilogram kelvin J/(kg·K)

SPECIFIC HEAT CAPACITY AT SATURATION c_{sat} Dim: $L^2T^{-2}\Theta^{-1}$
capacité thermique massique à saturation
spezifische Wärmekapazität bei Sättigung
SI unit: joule per kilogram kelvin J/(kg·K)

SPECIFIC HELMHOLTZ FREE ENERGY a, f Dim: L^2T^{-2}
énergie libre massique
spezifische freie Energie
SI unit: joule per kilogram J/kg
Note: also called 'specific Helmholtz function'

SPECIFIC IMPULSE I_{sp} Dim: T
impulsion spécifique, poussée spécifique
spezifischer Impuls, spezifischer Schub
SI unit: second s
Note: also called 'specific thrust'; parameter of rocket propellants

SPECIFIC INTERNAL ENERGY $u, (e)$ Dim: L^2T^{-2}
énergie interne massique
spezifische innere Energie
SI unit: joule per kilogram J/kg
M/S: MJ/kg, kJ/kg

SPECIFIC LATENT HEAT l Dim: L^2T^{-2}
chaleur latente massique
spezifische latente Wärme(menge)
SI unit: joule per kilogram J/kg
M/S: MJ/kg, kJ/kg

SPECIFIC LENGTH see: linear density

SPECIFIC RESISTANCE see: resistivity

SPECIFIC SURFACE see: surface density

SPECIFIC THRUST see: specific impulse

SPECIFIC VOLUME v Dim: L^3M^{-1}
volume massique
spezifisches Volumen
SI unit: cubic metre per kilogram m^3/kg
Note: the reciprocal is called 'density'

SPECIFIC WEIGHT γ Dim: $L^{-2}MT^{-2}$

poids volumique; poids spécifique

Wichte; spezifisches Gewicht

SI unit: newton per cubic metre N/m^3

Note: also called 'weight density'; obsolete, should be avoided and 'density' used instead

SPECTRAL ABSORPTION FACTOR $\alpha(\lambda)$ Dim: 1

facteur spectral d'absorption; absorptance spectrale

spektraler Absorptionsgrad

Note: also called 'spectral absorptance'

SPECTRAL ANGULAR CROSS SECTION $\sigma_{\Omega, E}$ Dim: $M^{-1}T^2$

section efficace différentielle spectrique; section efficace différentielle double

spektraler raumwinkelbezogener Wirkungsquerschnitt; (spektraler Winkelquerschnitt)

SI unit: square metre per steradian joule $m^2/(sr \cdot J)$

other unit: barn per steradian electronvolt $b/(sr \cdot eV)$

SPECTRAL CONCENTRATION OF RADIANT ENERGY DENSITY (in terms of wavelength) w_λ Dim: $L^{-2}MT^{-2}$

énergie rayonnante spectrique volumique (en longueur d'onde)

spektrale Strahlungsenergiedichte (wellenlängenbezogene)

SI unit: joule per metre to the fourth power J/m^4

Note: also called 'spectral radiant energy density (in terms of wavelength)'

SPECTRAL CROSS SECTION σ_E Dim: $M^{-1}T^2$

section efficace spectrique; section efficace différentielle énergétique

spektraler Wirkungsquerschnitt

SI unit: square metre per joule m^2/J

other unit: barn per electronvolt b/eV

SPECTRAL EMISSIVITY $\varepsilon(\lambda)$ Dim: 1

émissivité spectrale, émissivité à une longueur d'onde specifiée

spektraler Emissionsgrad

Note: also called 'emissivity at a specified wavelength'

SPECTRAL LUMINOUS EFFICACY $K(\lambda)$ Dim: $L^{-2}M^{-1}T^{3}J$

efficacité lumineuse spectrale, efficacité lumineuse à une longueur d'onde specifiée
spektrales photometrisches Strahlungsäquivalent
SI unit: lumen per watt lm/W
Note: also called 'luminous efficacy at a specified wavelength'

SPECTRAL LUMINOUS EFFICIENCY $V(\lambda)$ Dim: 1

efficacité lumineuse relative spectrale, efficacité lumineuse relative pour une longueur d'onde specifiée
spektraler Hellempfindlichkeitsgrad
Note: also called 'luminous efficiency at a specified wavelength'

SPECTRAL RADIAL ENERGY DENSITY see: spectral concentration . . .

SPECTRAL RADIANCE FACTOR $\beta(\lambda)$ Dim: 1

facteur spectral de luminance
spektraler Strahldichtefaktor

SPECTRAL REFLECTION FACTOR $\rho(\lambda)$ Dim: 1

facteur spectral de réflexion, réflectance spectrale
spektraler Reflexionsgrad
Note: also called 'spectral reflectance'

SPECTRAL TRANSMISSION FACTOR $\tau(\lambda)$ Dim: 1

facteur spectral de transmission, transmittance spectrale
spektraler Transmissionsgrad
Note: also called 'spectral transmittance'

SPEED see: velocity

SPIN ANGULAR MOMENTUM QUANTUM NUMBER s_i, S Dim: 1

nombre quantique du moment cinétique de spin
Spindrehimpuls-Quantenzahl

STANDARD ACCELERATION OF FREE FALL g_n Dim: LT^{-2}

valeur conventionnelle de l'accélération due à la pesanteur

Normfallbeschleunigung

$g_n =$ **9.806 65** m/s²

STANDARD AMBIENT CONDITIONS (—)

conditions normales ambiantes

Normalklima

temperature; **20** °C

relative humidity: **65%**

atmospheric pressure of dry air:

101 325 Pa = **1 013.25** mbar (= **1** atm)

STANDARD DENSITY ($\rho^{\ominus}$), (ρ_0) Dim: $L^{-3}M$

masse volumique normale

Normdichte

SI unit: kilogram per cubic metre kg/m³

Def: density of a substance at 'standard reference conditions' (q.v.)

STANDARD MOLAR VOLUME ($V_m^{\ominus}$) Dim: L^3N^{-1}

volume molaire normal

stoffmengenbezogenes (molares) Normvolumen

SI unit: cubic metre per mole m³/mol

Def: molar volume of a substance at 'standard reference conditions' (q.v.)

STANDARD MOLAR VOLUME OF AN IDEAL GAS $V_{m,0}$; ($V_{m,0}^{\ominus}$) Dim: L^3N^{-1}

volume molaire normal d'un gaz parfait

stoffmengenbezogenes (molares) Normvolumen des idealen Gases

SI unit: cubic metre per mole m³/mol

$V_{m,0} = 2.241\ 383 \times 10^{-2}$ m³/mol

Def: molar volume of an ideal gas at 'standard reference conditions' (q.v.)

STANDARD PRESSURE ($p^{\ominus}$); (p_0); (p_n) Dim: $L^{-1}MT^{-2}$

pression de référence, pression normale
Normdruck
SI unit: pascal Pa
oSI unit: bar bar
$p^{\ominus}$=**101 325** Pa=**1 013.25** mbar (=**1** atm)

STANDARD REFERENCE CONDITIONS (—)

conditions normales de référence
Normzustand
temperature: **0** °C
pressure: **101 325** Pa=**1 013.25** mbar (=**1** atm)
Note: also other values are used

STANDARD TEMPERATURE ($T^{\ominus}$, $t^{\ominus}$); (T_0, t_0); (T_n, t_n) Dim: Θ

température de référence, température normale
Normtemperatur
SI unit: kelvin K, degree Celsius °C
$T^{\ominus}$=**273.15** K, $t^{\ominus}$=**0** °C, ($T^{\ominus}=t^{\ominus}$)
Note: this is the commonly, but not exclusively, used value

STANDARD TEMPERATURE AND PRESSURE (STP, s.t.p.)

température et pression normales (TPN, PTN)
Normtemperatur und Normdruck (NTP, NPT)
Note: obsolescent; 'standard reference conditions' used instead

STANDARD VOLUME ($V^{\ominus}$); (V_0) Dim: L^3

volume normal
Normvolumen
SI unit: cubic metre m^3
Def: volume of a substance at 'standard reference conditions' (q.v.)

STANTON NUMBER *St* Dim: 1

nombre de Stanton
Stanton-Zahl
Note: also called 'Margoulis number' (q.v.)

STANTON NUMBER FOR MASS TRANSFER St^* Dim: 1
nombre de Stanton pour transfert de masse
Stanton-Zahl der Stoffübertragung

STATIC PRESSURE p_s Dim: $L^{-1}MT^{-2}$
pression statique
statischer Druck
SI unit: pascal Pa
M/S: mPa, μPa
other unit: bar bar

STATISTICAL WEIGHT g Dim: 1
poids statistique
statistisches Gewicht

STEFAN-BOLTZMANN CONSTANT σ Dim: $MT^{-3}\Theta^{-4}$
constante de Stefan-Boltzmann
Stefan-Boltzmann-Konstante
SI unit: watt per square metre kelvin to the fourth power $W/(m^2 \cdot K^4)$
$\sigma = 5.670\,32 \times 10^{-8}\ W/(m^2 \cdot K^4)$

STICKING PROBABILITY P_s Dim: 1
probabilité de collage, probabilité de sorption
Haftwahrscheinlichkeit

STOICHIOMETRIC NUMBER OF SUBSTANCE B ν_B Dim: 1
nombre stœchiométrique du constituant B
stöchiometrische Zahl eines Stoffes B

STOPPING POWER see: total linear s.p.

STRAIN see: linear s., shear s., volume s.

STRENGTH see: field s., ionic s.

STRESS see: normal s., shear s.

STROUHAL NUMBER Sr Dim: 1
nombre de Strouhal
Strouhal-Zahl

SUPERCONDUCTOR TRANSITION TEMPERATURE T_c Dim: Θ
température de transition supraconductrice
Supraleiterübergangstemperatur
SI unit: kelvin K

SURFACE COEFFICIENT OF HEAT TRANSFER h, α Dim: $MT^{-3}\Theta^{-1}$
coefficient de transmission thermique de surface
Wäremübergangskoeffizient
SI unit: watt per square metre kelvin $W/(m^2 \cdot K)$
Note: cf. 'coefficient of heat transfer'

SURFACE DENSITY ρ_A, (ρ_S) Dim: $L^{-2}M$
masse surfacique
flächenbezogene Masse, Massenbedeckung
SI unit: kilogram per square metre kg/m^2
Note: formerly called ('mass per unit area')

SURFACE DENSITY OF CHARGE σ Dim: $L^{-2}TI$
charge surfacique
Flächenladungsdichte, Ladungsbedeckung
SI unit: coulomb per square metre C/m^2
M/S: MC/m^2, C/mm^2, C/cm^2, kC/m^2, mC/m^2, $\mu C/m^2$

SURFACE TENSION γ, σ Dim: MT^{-2}
tension superficielle
Oberflächenspannung
SI unit: newton per metre N/m

SUSCEPTANCE B Dim: $L^{-2}M^{-1}T^3I^2$
susceptance
Blindleitwert, Suszeptanz
SI unit: siemens S
M/S: kS, mS, μS

T

TEMPERATURE see: Celsius t., Curie t., Debye t., Fermi t., Néel t., standard t., superconductor transition t., thermodynamic t.

TEMPERATURE DIFFERENCE ΔT, Δt Dim: Θ
différence de température
Temperaturdifferenz
SI unit: (for ΔT) kelvin K
(for Δt) degree Celsius °C
Note: ΔT is used for 'thermodynamic temperature difference', Δt is used for 'Celsius temperature difference'; also called 'temperature interval'

TENSION see: potential difference

THERMAL . . . see also 'heat . . .'

THERMAL COEFFICIENT OF LINEAR EXPANSION see: linear expansion c.

THERMAL CONDUCTANCE G Dim: $L^2MT^{-3}\Theta^{-1}$
conductance thermique
Wärmeleitwert
SI unit: watt per kelvin W/K
Note: the reciprocal is called 'thermal resistance'

THERMAL CONDUCTIVITY λ, (k) Dim: $LMT^{-3}\Theta^{-1}$
conductivité thermique
Wärmeleitfähigkeit
SI unit: watt per metre kelvin W/(m·K)

THERMAL DIFFUSION COEFFICIENT D_T Dim: L^2T^{-1}
coefficient de diffusion thermique
Thermodiffusionskoeffizient
SI unit: square metre per second m^2/s

THERMAL DIFFUSION FACTOR α_T Dim: 1
facteur de diffusion thermique
Thermodiffusionsfaktor

THERMAL DIFFUSION RATIO k_T Dim: 1
rapport de diffusion thermique
Thermodiffusionsverhältnis

THERMAL DIFFUSIVITY α, (α, κ) Dim: L^2T^{-1}
diffusivité thermique
Temperaturleitfähigkeit
SI unit: square metre per second m^2/s

THERMAL INSULANCE M Dim: $M^{-1}T^3\Theta$
coefficient d'isolation thermique
Wärmedurchgangswiderstand
SI unit: square metre kelvin per watt $m^2 \cdot K/W$
Note: also called 'coefficient of thermal insulation'

THERMAL RESISTANCE R Dim: $L^{-2}M^{-1}T^3\Theta$
résistance thermique
Wärmewiderstand
SI unit: kelvin per watt K/W
Note: the reciprocal of thermal resistance is called 'thermal conductance'

THERMAL RESISTIVITY (ρ_{th}) Dim: $L^{-1}M^{-1}T^3\Theta$
resistivité thermique
spezifischer Wärmewiderstand
SI unit: metre kelvin per watt $m \cdot K/W$
Note: the reciprocal of 'thermal conductivity'

THERMAL UTILIZATION FACTOR f Dim: 1
facteur d'utilisation thermique
thermische Nutzung

THERMODYNAMIC CRITICAL FIELD STRENGTH H_c Dim: $L^{-1}I$
intensité du champ critique thermodynamique
thermodynamische kritische Feldstärke
SI unit: ampere per metre A/m

THERMODYNAMIC TEMPERATURE T, Θ Dim: Θ
température thermodynamique; (formerly also *température absolue*)
thermodynamische Temperatur; (formerly also *absolute Temperatur*)
SI unit: kelvin K
Note: SI base quantity; formerly also called ('absolute temperature'). Cf. 'Celsius temperature' and 'temperature difference'

THERMOELECTROMOTIVE FORCE BETWEEN SUBSTANCES a AND b E_{ab} Dim: $L^2MT^{-3}I^{-1}$

force thermoélectromotrice entre deux substances a et b
thermoelektromotorische Kraft zwischen Stoffen a und b
SI unit: volt V

THICKNESS d, δ Dim: L

épaisseur
Dicke
SI unit: metre m
M/S: km, cm, mm, μm, nm, pm, fm

THOMSON COEFFICIENT μ, τ Dim: $L^2MT^{-3}I^{-1}\Theta^{-1}$

coefficient de Thomson
Thomson-Koeffizient
SI unit: volt per kelvin V/K

THROUGHPUT q_G Dim: L^2MT^{-3}

flux gazeux
pV-Durchfluß; (for a pump: *Saugleistung*)
SI unit: pascal cubic metre per second Pa·m³/s
oSI unit: pascal litre per second Pa·l/s, Pa·L/s
Note: flow rate in which the quantity of gas is expressed in pressure-volume units, hence the letters *pV* in the German name

TIME see: carrier life t., relaxation t., residence t., reverberation t.

TIME t Dim: T

temps
Zeit
SI unit: second s
M/S: ks, ms, μs, ns
oSI units: minute min, hour h, day d
Note: SI base quantity

TIME CONSTANT τ Dim: T

constante de temps
Zeitkonstante; Abklingzeit
SI unit: second s
Note: also called ‘relaxation time’; the reciprocal of the ‘damping coefficient’

TIME CONSTANT (of an exponentially varying quantity) τ, (T) Dim: T

constante de temps (*d'une grandeur variant exponentiellement*)
Zeitkonstante (*einer exponentiell sich ändernden Größe*)
SI unit: second s

TIME INTERVAL t Dim: T

intervalle de temps
Zeitspanne
SI unit: second s
Note: for M/S and oSI units see 'time'

TORQUE T Dim: L^2MT^{-2}

moment d'un couple; (*torque*)
Drehmoment
SI unit: newton metre N·m
Note: also called 'moment of a couple'

TOTAL ANGULAR MOMENTUM QUANTUM NUMBER j_i, J Dim: 1

nombre quantique du moment cinétique total
Gesamtdrehimpuls-Quantenzahl

TOTAL ATOMIC STOPPING POWER S_a Dim: L^4MT^{-2}

pouvoir d'arrêt atomique total
(*totales*) *atomares Bremsvermögen*
SI unit: joule square metre $J{\cdot}m^2$
oSI unit: electronvolt square metre $eV{\cdot}m^2$

TOTAL CROSS SECTION σ_{tot}, σ_T Dim: L^2

section efficace totale
totaler Wirkungsquerschnitt
SI unit: square metre m^2
other unit: barn b

TOTAL CROSS SECTION DENSITY see: total macroscopic c.s.

TOTAL IONIZATION BY A PARTICLE N_i Dim: 1

ionisation totale d'une particule
totale Ionisation eines Teilchens

TOTAL LINEAR STOPPING POWER S, S_l Dim: LMT^{-2}
pouvoir d'arrêt linéique total, (*pouvoir d'arrêt*)
(*totales*) *lineares Bremsvermögen*
SI unit: joule per metre J/m
oSI unit: electronvolt per metre eV/m
Note: also called ('stopping power')

TOTAL MACROSCOPIC CROSS SECTION Σ_{tot}, Σ_T Dim: L^{-1}
section efficace macroscopique totale, section efficace volumique totale
totaler makroskopischer Wirkungsquerschnitt, totale Wirkungsquerschnittsdichte
SI unit: reciprocal metre m^{-1}
Note: also called 'total cross section density'

TOTAL MASS STOPPING POWER S/ρ, (S_m) Dim: L^4T^{-2}
pouvoir d'arrêt massique total
(*totales*) *Massenbremsvermögen*
SI unit: joule square metre per kilogram $J \cdot m^2/kg$
oSI unit: electronvolt square metre per kilogram $eV \cdot m^2/kg$

TOTAL NEUTRON SOURCE DENSITY S Dim: $L^{-3}T^{-1}$
densité totale d'une source de neutrons
(*totale*) *Neutronenquelldichte*
SI unit: reciprocal second reciprocal cubic metre $s^{-1} \cdot m^{-3}$

TOTAL PRESSURE p Dim: $L^{-1}MT^{-2}$
pression totale
Totaldruck
SI unit: pascal Pa
oSI unit: bar bar
Note: the same as 'pressure'; sometimes used for the sum of partial pressures

TRANSMISSION COEFFICIENT τ Dim: 1
facteur de transmission
Schalltransmissionsgrad

TURNS RATIO n Dim: 1
rapport des nombres de spires
Windungszahlverhältnis; (Übersetzungsverhältnis)

U

UNDAMPED NATURAL FREQUENCY (f_0) Dim: T^{-1}
fréquence propre non amortie
Eigenfrequenz im ungedämpften Zustand, Kennfrequenz
SI unit: hertz Hz

UNITARY MASS DENSITY ρ_u Dim: $L^{-2}T^2$
masse volumique unitaire
druckbezogene Massendichte
SI unit: kilogram per cubic metre pascal kg/(m³·Pa)

UPPER CRITICAL FIELD STRENGTH H_{c2} Dim: $L^{-1}I$
intensité du champ critique supérieur
obere kritische Feldstärke
SI unit: ampere per metre A/m

V

VECTOR see: Burgers v., displacement v. . . ., lattice v., Poynting v.

VELOCITY see: angular v., sound particle v., volume v.

VELOCITY u, v, w, c Dim: LT^{-1}
vitesse
Geschwindigkeit
SI unit: metre per second m/s
oSI unit: kilometre per hour km/h
Note: also called 'linear velocity'

VELOCITY OF LIGHT c Dim: LT^{-1}
vitesse de la lumière
Lichtgeschwindigkeit
Note: see ‘velocity of propagation of electromagnetic waves’

VELOCITY OF PROPAGATION OF ELECTROMAGNETIC WAVES IN VACUUM c Dim: LT^{-1}
vitesse de propagation des ondes électromagnétiques dans le vide
Ausbreitungsgeschwindigkeit elektromagnetischer Wellen im Vakuum
SI unit: metre per second m/s
$c = 2.997\,924\,58 \times 10^8$ m/s
Note: sometimes c is used for velocity in a medium and c_0 for velocity in vacuum

VELOCITY OF SOUND c, (c_a) Dim: LT^{-1}
célérité, (*vitese du son*)
Schallgeschwindigkeit
SI unit: metre per second m/s
Note: see ‘Mach number’; cf. ‘sound particle velocity’

VISCOSITY see also: kinematic viscosity

VISCOSITY, DYNAMIC VISCOSITY η, (μ) Dim: $L^{-1}MT^{-1}$
viscosité; viscosité dynamique
dynamische Viskosität
SI unit: pascal second Pa·s
M/S: mPa·s

VOLTAGE see: potential difference

VOLUME see: molar v., specific v., standard molar v., standard v.

VOLUME V Dim: L^3
volume
Volumen, Rauminhalt
SI unit: cubic metre m^3
M/S: dm^3, cm^3, mm^3
oSI units: litre l, L (hl, hL.; cl, cL; ml, mL)

VOLUME COLLISION RATE χ Dim: $L^{-3}T^{-1}$
taux volumique de collision
volumenbezogene Stoßrate, Volumenstoßrate
SI unit: reciprocal cubic metre reciprocal second
$m^{-3} \cdot s^{-1}$

VOLUME DENSITY OF CHARGE ρ, (η) Dim: $L^{-3}TI$
charge volumique
Raumladungsdichte, Ladungsdichte
SI unit: coulomb per cubic metre C/m^3
M/S: C/mm^3, MC/m^3, C/cm^3, kC/m^3, mC/m^3, $\mu C/m^3$
Note: also called 'charge density'

VOLUME EXPANSION COEFFICIENT see: cubic e.c.

VOLUME FLOW RATE (of a fluid) q_V Dim: L^3T^{-1}
débit-volume
Volumenstrom, Volumendurchfluß, Volumendurchsatz
SI unit: cubic metre per second m^3/s

VOLUME FLOW RATE q, U Dim: L^3T^{-1}
flux de vitesse acoustique
Schallfluß
SI unit: cubic metre per second m^3/s
Note: this is a volume flow rate due to a sound wave; also called 'volume velocity'; often qualified e.g as 'instantaneous'

VOLUME FRACTION OF SUBSTANCE B ϕ_B Dim: 1
fraction volumique du constituant B
Volumenanteil (Volumengehalt) eines Stoffes B; (Volumenbruch eines Stoffes B)
Note: usually expressed in %, ‰, or ppm; volume fraction in % is intended to replace '% by volume' and incorrect 'volume %'.

VOLUME PERCENT (vol. %, % vol.)
pour cent en volume (% vol.)
Volumenprozent (Vol.-%)
Note: also called 'percent by volume' (% by vol.); see 'volume fraction'

VOLUME STRAIN θ Dim: 1

dilatation volumique relative
relative Volumenänderung
Note: also called ('bulk strain')

VOLUME VELOCITY see: volume flow rate

W

WAVELENGTH see: Compton w.

WAVELENGTH λ Dim: L

longueur d'onde
Wellenlänge
SI unit: metre m
M/S: mm, µm, nm, pm
Note: the reciprocal is called 'wavenumber'

WAVENUMBER see: circular w., Debye circular w., Fermi circular w.

WAVENUMBER σ Dim: L^{-1}

nombre d'onde (*linéique*)
Wellenzahl; Repetenz
SI unit: reciprocal metre m^{-1}
Note: also called 'repetency'

WEBER NUMBER *We* Dim: 1

nombre de Weber
Weber-Zahl

WEIGHT see also: atomic w., molecular w., specific w., statistical w.

WEIGHT *G*, (*P*, *W*) Dim: LMT^{-2}

poids
Gewichtskraft; (*Gewicht*)
SI unit: newton N
Note: this quantity should be avoided and 'mass' used instead

WEIGHT PERCENT (wt %)

pour cent en poids (% en poids)

Gewichtsprozent (Gew.-%)

Note: deprecated; also called 'percent by weight' (% by wt); see 'mass fraction'

WIDTH see: level w., also see: breadth

WORK $W, (A)$ Dim: L^2MT^{-2}

travail

Arbeit

SI unit: joule J

M/S: EJ, PJ, TJ, GJ, MJ, kJ, mJ

WORK FUNCTION Φ Dim: L^2MT^{-2}

travail d'extraction; travail de sortie

Austrittsarbeit

SI unit: joule J

oSI unit: electronvolt eV

Y

YOUNG MODULUS see: modulus of elasticity

* * * *

η-FACTOR see: neutron yield per absorption

ν-FACTOR see: neutron yield per fission

Part 3

Symbols Denoting Quantities and Constants

A	activity, affinity, area, equivalent absorption area, Helmholtz free energy, linear current density, magnetic vector potential, nucleon number, permeance, Richardson constant, work
A_{H}	Hall coefficient
A_{r}	relative atomic mass
Al	Alfvén number
a	acceleration, linear absorption coefficient, sound particle acceleration, specific activity, specific Helmholtz free energy, thermal diffusivity
a_{A}	activity of solvent substance A
a_{B}	activity of solute substance B
a_m	mass absorption coefficient
$a_{m,\,\mathrm{B}}$	activity of solute substance B
a_0	Bohr radius
B	magnetic flux density, mass defect, susceptance
B_{i}	magnetic polarization
B_{r}	relative mass defect
Bi	Biot number
b	binding fraction, breadth, mobility ratio
$\boldsymbol{b}$	Burgers vector
b_{B}	molality of solute substance B
C	capacitance, conductance, heat capacity
C_{B}	molecular concentration of substance B
C_{i}	intrinsic conductance
C_{m}	molar heat capacity
C_{N}	molecule conductance
Co	Cowling number
c	specific heat capacity, velocity of propagation of electromagnetic waves (in vacuum), velocity of sound
c_{a}	velocity of sound

c_B concentration of substance B

c_p specific heat capacity at constant pressure

c_{sat} specific heat capacity at saturation

c_V specific heat capacity at constant volume

c_0 velocity of propagation of electromagnetic waves in vacuum

c_1 first radiation constant

c_2 second radiation constant

D absorbed dose, Debye-Waller factor, diameter, diffusion coefficient, diffusion coefficient for neutron fluence rate, diffusion coefficient (for neutron number density), electric flux density, power of a lens

$\dot{D}$ absorbed dose rate

D_i electric polarization

D_n diffusion coefficient

D_T thermal diffusion coefficient

D_ϕ diffusion coefficient for neutron fluence rate

d diameter, dissipation factor, relative density, thickness

$d_{\frac{1}{2}}$ half-thickness

E electric field strength, electromotive force, energy, illuminance, internal energy, irradiance, modulus of elasticity, sound energy density

E_a acceptor ionization energy

E_{ab} thermoelectromotive force between substances a and b

E_d donor ionization energy

E_e irradiance

E_F Fermi energy

E_g gap energy

E_h Hartree energy

E_k kinetic energy

E_m	molar internal energy
E_p	potential energy
E_r	resonance energy
E_{res}	resonance energy
E_v	illuminance
E_β	maximum beta particle energy
Eu	Euler number
e	elementary charge, linear strain, specific internal energy
F	Faraday constant, force, Helmholtz free energy, hyperfine structure quantum number, magnetomotive force
F_m	magnetomotive force
Fo	Fourier number
*Fo**	Fourier number for mass transfer
Fr	Froude number
f	coefficient of friction, frequency, gravitational constant, packing fraction, specific Helmholtz free energy, thermal utilization factor
f_B	activity coefficient of substance B, fugacity of substance B
f_d	damped natural frequency
f_0	undamped natural frequency
G	conductance, Gibbs free energy, gravitational constant, quantity of gas, shear modulus, thermal conductance, weight
Gr	Grashof number
*Gr**	Grashof number for mass transfer
g	acceleration of free fall, *g*-factor . . ., local acceleration of free fall, specific Gibbs free energy, statistical weight
g_n	standard acceleration of free fall

H	dose equivalent, enthalpy, light exposure, magnetic field strength, radiant exposure, action
H_c	thermodynamic critical field strength
H_{c1}	lower critical field strength
H_{c2}	upper critical field strength
H_e	radiant exposure
H_i	magnetization
Ha	Hartmann number
h	coefficient of heat transfer, depth, height, Planck constant, specific enthalpy, surface coefficient of heat transfer
$\hbar$	Dirac constant
I	electric current, enthalpy, ionic strength, luminous intensity, moment of inertia, nuclear spin quantum number, radiant intensity, second (axial) moment of area, sound intensity
I_a	second (axial) moment of area
I_e	radiant intensity
I_p	second polar moment of area
I_{sp}	specific impulse
I_v	luminous intensity
i	specific enthalpy
i_f	frequency interval
J	current density, current density of particles, exchange integral, ion dose, magnetic polarization, Massieu function, moment of inertia, sound intensity, total angular momentum quantum number
j	heat transfer factor, ion dose rate, magnetic dipole moment
j_i	total angular momentum quantum number
j_m	mass transfer factor
K	bulk modulus, coefficient of heat transfer, electric field strength, equilibrium constant, kerma, kinetic energy, luminous efficacy

$\dot{K}$	kerma rate
K_{m}	maximum spectral luminous efficacy
$K(\lambda)$	spectral luminous efficacy
Kn	Knudsen number
k	Boltzmann constant, circular wavenumber, coefficient of heat transfer, coupling coefficient, multiplication factor
k_{eff}	effective multiplication factor
k_{F}	Fermi circular wavenumber
k_{T}	thermal diffusion ratio
k_{∞}	infinite medium multiplication factor
L	Avogadro constant, diffusion length, latent heat, length, linear energy transfer, Lorenz coefficient, luminance, moment of momentum, orbital angular momentum quantum number, radiance, self inductance
L_{e}	radiance
L_{I}	sound intensity level
L_{N}	loudness level
L_{n}	diffusion length
L_{P}	sound power level
L_{p}	diffusion length
L_{p}	sound pressure level
L_{s}	slowing-down length
L_{s}^{2}	slowing-down area
L_{sl}	slowing-down length
L_{sl}^{2}	slowing-down area
L_{v}	luminance
L_{W}	sound power level
L_{12}	mutual inductance
L^{2}	diffusion area
Le	Lewis number

l	attenuation length, length, mean free path, mean free path of electrons, specific latent heat
l_e	mean free path of electrons
l_i	orbital angular momentum quantum number
l_{ph}	mean free path of phonons
M	luminous exitance, magnetic quantum number, magnetization, migration length, molar mass, moment of force, mutual inductance, radiant exitance, thermal insulance
M^2	migration area
M_e	radiant exitance
M_r	relative molecular mass
M_v	luminous exitance
Ma	Mach number
Ms	Margoulis number
m	electromagnetic moment, mass, mass of molecule, number of phases
m_a	mass of atom
m_B	molality of solute substance B
m_e	rest mass of electron
m_i	magnetic quantum number
m_n	rest mass of neutron
m_p	rest mass of proton
m_u	atomic mass constant (unified)
$m(X)$	mass of atom (of a nuclide X)
m^*	effective mass
N	loudness, neutron number, number of molecules (or particles), number of turns in a winding
N_A	Avogadro constant
N_a	acceptor number density
N_d	donor number density
N_E	density of states

N_i	total ionization by a particle
$N_{\mathrm{i}l}$	linear ionization by a particle
N_L	Loschmidt constant
Nu	Nusselt number
Nu^*	Nusselt number for mass transfer
n	amount of substance, electron number density, neutron number density, number density of molecules (or particles), order of reflexion, principal quantum number, refractive index, rotational frequency, turns ratio
n_a	acceptor number density
n_d	donor number density
n_n	electron number density
n_p	hole number density
n^-	ion number density
n_-	electron number density
n^+	ion number density
n_+	hole number density
P	active power, electric polarization, permeance, power, radiant power, sound energy flux, weight
P_a	sound energy flux
P_q	reactive power
P_s	apparent power, sticking probability
Pe	Péclet number
Pe^*	Péclet number for mass transfer
Pr	Prandtl number
p	electric dipole moment, electric dipole moment of molecule, hole number density, momentum, number of pairs of poles, pressure, resonance escape probability, sound pressure, total pressure
p_a	sound pressure
p_abs	absolute pressure
p_amb	ambient pressure

$\bar{p}_B$	fugacity of substance B
p_e	electric dipole moment, gauge pressure
p_L	saturation vapour pressure
p_n	standard pressure
p_s	static pressure
p_0	standard pressure
$p^{\ominus}$	standard pressure
Q	canonical partition function, electric charge, heat, nuclear quadrupole moment, quantity of light, radiant energy, reaction energy, reactive power
Q_e	radiant energy
Q_v	quantity of light
Q_α	alpha disintegration energy
Q_β	beta disintegration energy
q	circular wavenumber, density of heat flow rate, molecular partition function, slowing-down density, volume flow rate
q_D	Debye circular wavenumber
q_G	throughput
q_m	mass flow rate
q_N	molecule flow rate
q_V	volume flow rate
q_ν	molar flow rate
R	mean linear range, molar gas constant, nuclear radius, reluctance, resistance, sound reduction index, thermal resistance
$\boldsymbol{R}$	lattice vector
R_H	Hall coefficient
R_l	mean linear range
R_{m}	reluctance
R_m	mean mass range
$\boldsymbol{R}_0$	lattice vector

R_∞	Rydberg constant
R_ρ	mean mass range
Ra	Rayleigh number
Re	Reynolds number
Rm	magnetic Reynolds number
Ry	Rydberg energy
r	jerk, radius, reflection coefficient
r_B	mole ratio of solute substance B
r_e	electron radius
S	apparent power, area, current density, current density of particles, entropy, Poynting vector, spin angular momentum quantum number, total linear stopping power, total neutron source density
S_a	total atomic stopping power
S_ab	Seebeck coefficient for substances a and b
S_l	total linear stopping power
$S_{l,\mathrm{r}}$	relative linear stopping power
S_m	molar entropy
S_m	total mass stopping power
$S_{m,\mathrm{r}}$	relative mass stopping power
Sc	Schmidt number
Sh	Sherwood number
Sr	Strouhal number
St	Stanton number
St^*	Stanton number for mass transfer
s	length of path, long range order parameter, specific entropy
s_i	spin angular momentum quantum number
T	kinetic energy, period, reactor time constant, reverberation time, thermodynamic temperature, torque
$\boldsymbol{T}$	lattice vector
T_C	Curie temperature

T_c	superconductor transition temperature
T_F	Fermi temperature
T_N	Néel temperature
T_n	standard temperature
$T^\ominus$	standard temperature
T_0	standard temperature
$T_{\frac{1}{2}}$	half-life
t	Celsius temperature, duration, time, time interval
t_n	standard temperature
t_0	standard temperature
$t^\ominus$	standard temperature
U	conductance, internal energy, magnetic potential difference, potential difference, radiant energy, volume flow rate
U_i	intrinsic conductance
U_m	magnetic potential difference, molar internal energy
U_N	molecule conductance
u	lethargy, radiant energy density, sound particle velocity, velocity
$\boldsymbol{u}$	displacement vector of ion
V	electric potential, luminous efficiency, potential difference, potential energy, volume
V_m	molar volume
$\mathrm{V}_m^\ominus$	standard molar volume
$V_{m,0}$	standard molar volume of an ideal gas
$V_{m,0}^\ominus$	standard molar volume of an ideal gas
V_0	standard volume
$V(\lambda)$	spectral luminous efficiency
$V^\ominus$	standard volume
v	impingement rate, neutron speed, sound particle velocity, specific volume, velocity

W	energy, radiant energy, section modulus, weight, work
W_i	average energy loss per ion pair formed
We	Weber number
w	electromagnetic energy density, energy density, radiant energy density, resistance, sound energy density, velocity
w_a	sound energy density
w_B	mass fraction of substance B
w_λ	spectral concentration of radiant energy density

X	exposure, reactance
$\dot{X}$	exposure rate
x	chromaticity co-ordinate
x_B	mole fraction of substance B
$\bar{x}(\lambda)$	CIE spectral tristimulus value

Y	admittance, Planck function
$\|Y\|$	modulus of admittance
y	chromaticity co-ordinate
y_B	mole fraction of substance B
$\bar{y}(\lambda)$	CIE spectral tristimulus value

Z	atomic number, canonical partition function, impedance, proton number
$\|Z\|$	modulus of impedance
Z_a	acoustic impedance
Z_c	characteristic impedance of a medium
Z_m	mechanical impedance
Z_s	specific acoustic impedance
z	charge number of ion, chromaticity co-ordinate
$\bar{z}(\lambda)$	CIE spectral tristimulus value

α	acoustic absorption coefficient, angle (plane), angular acceleration, attenuation coefficient, coefficient of heat transfer, degree of dissociation, electric polarizability of molecule, fine-structure constant, internal conversion factor, linear current density, Madelung constant, recombination coefficient, surface coefficient of heat transfer, thermal diffusivity
α_a	acoustic absorption coefficient
α_l	linear expansion coefficient
α_p	relative pressure coefficient
α_T	thermal diffusion factor
α_V	cubic expansion coefficient
$\alpha(\lambda)$	spectral absorption factor
β	angle (plane), phase coefficient, pressure coefficient,
$\beta(\lambda)$	spectral radiance factor
Γ	Grüneisen parameter, level width, specific gamma ray constant
γ	angle (plane), conductivity, cubic expansion coefficient, Grüneisen parameter, gyromagnetic coefficient, propagation coefficient, ratio of specific heat capacities, shear strain, specific heat, surface tension
γ_B	activity coefficient of solute substance B
Δ	mass excess
Δp	differential pressure
$\mathbf{\Delta}_r$	relative mass excess
Δ_T	temperature difference
Δ_t	temperature difference
δ	damping coefficient, dissipation coefficient, loss angle, thickness
ε	emissivity, energy imparted, fast fission factor, linear strain, permittivity
ε_{ab}	Seebeck coefficient for substances a and b

ε_F	Fermi energy
ε_r	relative permittivity
ε_0	permittivity of vacuum
$\varepsilon(\lambda)$	spectral emissivity
$\varepsilon(\lambda, \theta, \phi)$	directional spectral emissivity
η	efficiency, neutron yield per absorption, viscosity, volume density of charge
Θ	current linkage, thermodynamic temperature
Θ_D	Debye temperature
θ	angle (plane), Bragg angle, Celsius temperature, volume strain
κ	compressibility, coupling coefficient, electrolytic conductivity, isentropic exponent, Landau-Ginzburg parameter, magnetic susceptibility, molar absorption coefficient, thermal diffusivity
Λ	logarithmic decrement, mean free path of phonons, non-leakage probability, permeance
Λ_m	molar conductivity
λ	decay constant, mean free path, power factor, thermal conductivity, wavelength
λ_B	absolute activity of substance B
λ_C	Compton wavelength
λ_L	London penetration depth
μ	coefficient of friction, electric dipole moment of molecule, linear attentuation coefficient, magnetic moment of a particle, mobility, permeability, Poisson ratio, Thomson coefficient, viscosity
μ/ρ	mass attenuation coefficient
μ_a	atomic attenuation coefficient
μ_{at}	atomic attenuation coefficient
μ_B	Bohr magneton, chemical potential of substance B
μ_{en}/ρ	mass energy absorption coefficient

μ_l	linear attentuation coefficient
μ_m	mass attenuation coefficient
μ_N	nuclear magneton
μ_r	relative permeability
μ_{tr}/ρ	mass energy transfer coefficient
μ_0	permeability of vacuum
ν	amount of substance, frequency, kinematic viscosity, neutron yield per fission, Poisson ratio
ν_B	stoichiometric number of substance B
ν_c	cyclotron frequency
ν_L	Larmor frequency
Ξ	grand-canonical partition function
ξ	average logarithmic energy decrement, coherence length, sound particle displacement
Π	osmotic pressure
Π_{ab}	Peltier coefficient for substances a and b
ρ	density, density of states, reactivity reflection coefficient, resistivity, volume density of charge
ρ_A	surface density
ρ_B	mass concentration of substance B, partial pressure of substance B
ρ_l	linear density
ρ_S	surface density
ρ_u	unitary mass density
ρ_0	standard density
$\rho(\lambda)$	spectral reflection factor
$\rho^\ominus$	standard density
Σ	macroscopic cross section
Σ_T	total macroscopic cross section
Σ_{tot}	total macroscopic cross section

σ conductivity, cross section, leakage coefficient, normal stress, short range order parameter, Stefan-Boltzmann constant, surface density of charge, wavenumber

σ_A absorption cross section

σ_a absorption cross section

σ_E spectral cross section

σ_f fission cross section

σ_S scattering cross section

σ_s scattering cross section

σ_T total cross section

σ_{tot} total cross section

σ_Ω angular section

$\sigma_{\Omega,E}$ spectral angular cross section

τ carrier life time, mean life, relaxation time, residence time, shear stress, Thomson coefficient, time constant, time constant (of an exponentially varying quantity), transmission coefficient

τ_n carrier life time

τ_p carrier life time

$\tau(\lambda)$ spectral transmission factor

Φ heat flow rate, luminous flux, magnetic flux, particle fluence, potential energy, radiant power, work function

Φ_e radiant power

Φ_v luminous flux

Φ_0 fluxoid quantum

ϕ angle (plane), density of heat flow rate, electric potential, neutron fluence rate, osmotic coefficient of solvent substance A, particle fluence rate, phase difference, radiant energy fluence rate, fluidity

ϕ_B volume fraction of substance B

χ	electric susceptibility, volume collision rate
χ_e	electric susceptibility
Ψ	energy fluence, electric flux
ψ	collision rate, dissipation coefficient, energy fluence rate, radiant energy fluence rate
Ω	microcanonical partition function, solid angle
ω	angular velocity, circular frequency
ω_c	cyclotron angular frequency
ω_D	Debye circular frequency
ω_L	Larmor angular frequency
ω_N	nuclear precession angular frequency

NOTE

Because italic Greek alpha, 'α', and nu, 'ν', are almost indistinguishable from italic '*a*' and '*v*', roman alpha, 'α', and nu, 'ν', have been used in all cases where these symbols are used. While this is not in accordance with ISO convention, it avoids any possible confusion.

Appendix 1: Definitions of Base SI Units

THE METRE is the length equal to 1 650 763.73 wavelengths in vacuum of the radiation corresponding to the transition between the levels $2p_{10}$ and $5d_5$ of the krypton-86 atom (11th CGPM, 1960).

THE KILOGRAM is the unit of mass; it is equal to the mass of the international prototype of the kilogram (1st CGPM, 1889 and 3rd CGPM, 1901).

THE SECOND is the duration of 9 192 631 770 periods of the radiation corresponding to the transition between the two hyperfine levels of the ground state of the caesium–133 atom (13th CGPM, 1967).

THE AMPERE is that constant electric current which, if maintained in two straight parallel conductors of infinite length, of negligible circular cross-section, and placed 1 metre apart in vacuum, would produce between these conductors a force equal to 2×10^{-7} newton per metre of length (CIPM, 1946 and 9th CGPM, 1948).

THE KELVIN, unit of thermodynamic temperature, is the fraction 1/273.16 of the thermodynamic temperature of the triple point of water (13th CGPM, 1967).

THE MOLE is the amount of substance of a system which contains as many elementary entities as there are atoms in 0.012 kilogram of carbon-12. When the mole is used the elementary entities must be specified and may be atoms, molecules, ions, electrons, other particles, or specified groups of such particles (14th CGPM, 1971).
Note: in this definition it is understood that unbound atoms of carbon 12, at rest and in their ground state, are referred to (CIPM, 1980).

THE CANDELA is the luminous intensity, in a given direction, of a source that emits monochromatic radiation of frequency 540×10^{12} hertz and that has a radiant intensity in that direction of (1/683) watt per steradian (16th CGPM, 1979).

Appendix 2: UK and US Units

Table 1. UK and US units of length.

	in	ft	yd	chain	furlong	mile
in	1	1/12	1/36	1/792	1/7920	1/63 360
ft	12	1	1/3	1/66	1/660	1/5280
yd	36	3	1	1/22	1/220	1/1760
chain	792	66	22	1	1/10	1/80
furlong	7920	660	220	10	1	1/8
mile	63 360	5280	1760	80	8	1

Table 2. UK and US units of area.

	in^2	ft^2	yd^2	rood	acre	$mile^2$
in^2	1	1/144	1/1296	–	–	–
ft^2	144	1	1/9	1/10 890	1/43 560	–
yd^2	1296	9	1	1/1210	1/4840	$3.228\,31 \times 10^{-7}$
rood	–	10 890	1210	1	1/4	1/2560
acre	–	43 560	4840	4	1	1/640
$mile^2$	–	–	$3.097\,6 \times 10^{6}$	2560	640	1

Table 3. UK and US units of volume.

	in^3	ft^3	yd^3
in^3	1	1/1728	1/46 656
ft^3	1728	1	1/27
yd^3	46 656	27	1
$mile^3$	$2.543\,58 \times 10^{14}$	$1.471\,98 \times 10^{11}$	$5.451\,78 \times 10^{9}$

Table 4. UK units of volume (capacity).

bushel	peck	gallon	quart	pint	gill	fl oz	fl dr	minim
1	4	8	32	64	256	1280	10240	614400
1/4	1	2	8	16	64	320	2560	153600
1/8	1/2	1	4	8	32	160	1280	76800
1/32	1/8	1/4	1	2	8	40	320	19200
1/64	1/16	1/8	1/2	1	4	20	160	9600
1/256	1/64	1/32	1/8	1/4	1	5	40	2400
1/1280	1/320	1/160	1/40	1/20	1/5	1	8	480
1/10240	1/2560	1/1280	1/320	1/160	1/40	1/8	1	60
1/614400	1/153600	1/76800	1/19200	1/9600	1/2400	1/480	1/60	1

Table 5. US units of volume (capacity)—for liquid measure only.

	gallon	lq qt	lq pt	gill	fl oz	fl dr	minim
gallon	1	4	8	32	128	1024	61440
lq quart	1/4	1	2	8	32	256	15360
lq pint	1/8	1/2	1	4	16	128	7680
gill	1/32	1/8	1/4	1	4	32	1920
fl oz	1/128	1/32	1/16	1/4	1	8	480
fl dr	1/1024	1/256	1/128	1/32	1/8	1	60
minim	1/61440	1/15360	1/7680	1/1920	1/480	1/60	1

Table 6. US units of volume (capacity)—dry measure only.

	bu	pk	dry qt	dry pt
bushel	1	4	32	64
peck	1/4	1	8	16
dry quart	1/32	1/8	1	2
dry pint	1/64	1/16	1/2	1

Table 7. UK units of mass.

	ton	cwt	(qr)	stone	lb	oz	dr
ton	1	20	80	160	2240	35 840	573 440
hundredweight	1/20	1	4	8	112	1792	28 672
quarter	1/80	1/4	1	2	28	448	7168
stone	1/160	1/8	1/2	1	14	224	3584
pound (avoir)	1/2240	1/112	1/28	1/14	1	16	256
ounce (avoir)	1/35 840	1/1792	1/448	1/224	1/16	1	16
dram (avoir)	1/573 440	1/28 672	1/7168	1/3584	1/256	1/16	1

Table 8. US units of mass.

	(sh ton)	(sh cwt)	lb	oz	dr
short ton	1	20	2000	32 000	512 000
short hundredweight	1/20	1	100	1600	25 600
pound (avoir)	1/2000	1/100	1	16	256
ounce (avoir)	1/32 000	1/1600	1/16	1	16
dram (avoir)	1/512 000	1/125 600	1/256	1/16	1

Table 9. Troy units of mass.

	lb tr	oz tr	dwt	gr
troy pound	1	12	240	5760
troy ounce	1/12	1	20	480
pennyweight	1/240	1/20	1	24
grain	1/5760	1/480	1/24	1

Table 10. Apothecaries' units of mass.

	oz apoth	dr ap	scruple	gr
apoth. ounce	1	8	24	480
drachm; dram	1/8	1	3	60
scruple	1/24	1/3	1	20
grain	1/480	1/60	1/20	1

Appendix 3: Recommended Consistent Values of the Fundamental Constants

Quantity	*Symbol*	*Value*	*Uncertainty* (ppm)
1. Permeability of Vacuum	μ_0	$4\pi \times 10^{-7}$ H m^{-1} $= 12.566\,370\,614\,4 \times 10^{-7}$ H m^{-1}	
2. Speed of Light in Vacuum	c	$2.997\,924\,58(1.2) \times 10^{8}$ m s^{-1}	0.004
3. Permittivity of Vacuum	$\varepsilon_0 = (\mu_0 c^2)^{-1}$	$8.854\,187\,82(7) \times 10^{-12}$ F m^{-1}	0.008
4. Fine Structure Constant, $\mu_0 ce^2/2h$	α	0.007 297 350 6(60)	0.82
	α^{-1}	137.036 04(11)	0.82
5. Elementary Charge	e	$1.602\,189\,2(46) \times 10^{-19}$ C	2.9
6. Planck Constant	h	$6.626\,176(36) \times 10^{-34}$ J Hz^{-1}	5.4
	$\hbar = h/2\pi$	$1.054\,588\,7(57) \times 10^{-34}$ J s	5.4
7. Avogadro Constant	N_A	$6.022\,045(31) \times 10^{23}$ mol^{-1}	5.1
8. Atomic Mass Unit	$1\,u = (10^{-3}$ kg mol$^{-1})/N_A$	$1.660\,565\,5(86) \times 10^{-27}$ kg	5.1
9. Electron Rest Mass	m_e	$0.910\,953\,4(47) \times 10^{-30}$ kg	5.1
		$5.485\,802\,6(21) \times 10^{-4}$ u	0.38
10. Muon Rest Mass	m_μ	$1.883\,566(11) \times 10^{-28}$ kg	5.6
		0.113 429 20(26) u	2.3
11. Proton Rest Mass	m_p	$1.672\,648\,5(86) \times 10^{-27}$ kg	5.1
		1.007 276 470(11) u	0.011
12. Neutron Rest Mass	m_n	$1.674\,954\,3(86) \times 10^{-27}$ kg	5.1
		1.008 665 012(37) u	0.037
13. Ratio, Proton Mass to Electron Mass	m_p/m_e	1836.151 52(70)	0.38
14. Ratio, Muon Mass to Electron Mass	m_μ/m_e	206.768 65(47)	2.3

15. Specific Electron Charge	e/m_e	$1.758\,804\,7(49)\times10^{11}$ C kg^{-1}	2.8
16. Faraday Constant	$F=N_Ae$	$9.648\,456(27)\times10^{4}$ C mol^{-1}	2.8
17. Magnetic Flux Quantum	$\Phi_0=h/2e$	$2.067\,850\,6(54)\times10^{-15}$ Wb	2.6
	h/e	$4.135\,701(11)\times10^{-15}$ J Hz^{-1} C^{-1}	2.6
18. Josephson Frequency-Voltage Ratio	$2e/h$	483.593 9(13) THz V^{-1}	2.6
19. Quantum of Circulation	$h/2m_e$	$3.636\,945\,5(60)\times10^{-4}$ J Hz^{-1} kg^{-1}	1.6
	h/m_e	$7.273\,891(12)\times10^{-4}$ J Hz^{-1} kg^{-1}	1.6
20. Rydberg Constant	R_∞	$1.097\,373\,177(83)\times10^{7}$ m^{-1}	0.075
21. Bohr Radius	$a_0=\alpha/4\pi R_\infty$	$0.529\,177\,06(44)\times10^{-10}$ m	0.82
22. Electron Compton Wavelength	$\lambda_C=\alpha^2/2R_\infty$	$2.426\,308\,9(40)\times10^{-12}$ m	1.6
	$\lambda\!\!\!^{-}_C=\lambda_C/2\pi=\alpha a_0$	$3.861\,590\,5(64)\times10^{-13}$ m	1.6
23. Classical Electron Radius	$r_e=\mu_0e^2/4\pi m_e=\alpha\lambda\!\!\!^{-}_C$	$2.817\,938\,0(70)\times10^{-15}$ m	2.5
24. Electron g-Factor	$\frac{1}{2}g_e=\mu_e/\mu_B$	1.001 159 656 7(35)	0.0035
25. Muon g-Factor	$\frac{1}{2}g_\mu$	1.001 166 16(31)	0.31
26. Proton Moment in Nuclear Magnetons	μ_p/μ_N	2.792 845 6(11)	0.38
27. Bohr Magneton	$\mu_B=e\hbar/2m_e$	$9.274\,078(36)\times10^{-24}$ J T^{-1}	3.9
28. Nuclear Magneton	$\mu_N=e\hbar/2m_p$	$5.050\,824(20)\times10^{-27}$ J T^{-1}	3.9
29. Electron Magnetic Moment	μ_e	$9.284\,832(36)\times10^{-24}$ J T^{-1}	3.9
30. Proton Magnetic Moment	μ_p	$1.410\,617\,1(55)\times10^{-26}$ J T^{-1}	3.9
31. Proton Magnetic Moment in Bohr Magnetons	μ_p/μ_B	$1.521\,032\,209(16)\times10^{-3}$	0.011
32. Ratio, Electron to Proton Magnetic Moments	μ_e/μ_p	658.210 6880 (66)	0.010
33. Ratio, Muon Moment to Proton Moment	μ_μ/μ_p	3.183 340 2(72)	2.3
34. Muon Magnetic Moment	μ_μ	$4.490\,474(18)\times10^{-26}$ J T^{-1}	3.9
35. Proton Gyromagnetic Ratio	γ_p	$2.675\,198\,7(75)\times10^{8}$ s^{-1} T^{-1}	2.8
36. Diamagnetic Shielding Factor, Spherical H_2O Sample	$1+\sigma(H_2O)$	1.000 025 637(67)	0.067
37. Proton Gyromagnetic Ratio, (Uncorrected)	γ_p'	$2.675\,130\,1(75)\times10^{8}$ s^{-1}T^{-1}	2.8
	$\gamma_p'/2\pi$	42.576 02(12) MHz T^{-1}	2.8
38. Proton Moment in Nuclear Magnetons (Uncorrected)	μ_p'/μ_N	2.792 774 0(11)	0.38

Appendix 3: Continued:—

Quantity	Symbol	Value	Uncertainty (ppm)
39. Proton Compton Wavelength	$\lambda_{C,p}=h/m_pC$	$1.321\,409\,9(22)\times 10^{-15}$ m	1.7
	$\bar{\lambda}_{C,p}=\lambda_{C,p}/2\pi$	$2.103\,089\,2(36)\times 10^{-16}$ m	1.7
40. Neutron Compton Wavelength	$\lambda_{C,n}=h/m_nC$	$1.319\,590\,9(22)\times 10^{-15}$ m	1.7
	$\bar{\lambda}_{C,n}=\lambda_{C,n}/2\pi$	$2.100\,194\,1(35)\times 10^{-16}$ m	1.7
41. Molar Gas Constant	R	8.314 41(26) J mol^{-1} K^{-1}	31
42. Molar Volume, Ideal Gas ($T_0=273.15$ K, $p_0=1$ atm)	$V_m=RT_0/p_0$	0.022 413 83(70) m^3 mol^{-1}	31
43. Boltzmann Constant	$k=R/N_A$	$1.380\,662(44)\times 10^{-23}$ J K^{-1}	32
44. Stefan-Boltzmann Constant	$\sigma=(\pi^2/60)k^4/\hbar^3c^2$	$5.670\,32(71)\times 10^{-8}$ W m^{-2} K^{-4}	125
45. First Radiation Constant	$c_1=2\pi hc^2$	$3.741\,832(20)\times 10^{-16}$ W m^2	5.4
46. Second Radiation Constant	$c_2=hc/k$	0.014 387 86(45) m K	31
47. Gravitational Constant	G	$6.672\,0(41)\times 10^{-11}$ N m^2 kg^{-2}	615

Appendix 4: Bibliography

Australian Standards (AS)

1000:1979 The International System of Units (SI) and its application

1376:1973 Conversion factors

Austrian Standards (ÖNORM)

A 6401	09.76	Zeichen für Größen und Einheiten
A 6432	05.77	Internationales Einheitensystem (SI)
A 6433	09.78	Geometrie; Größen und Einheiten
A 6434	12.78	Kinematik; Größen und Einheiten
A 6435	06.80	Mechanik; Größen und Einheiten
A 6436	03.81	Wärme; Größen und Einheiten
A 6437	11.73	Maßsysteme; Licht und andere optische Strahlungen

Belgian Standards (NBN)

X 02–001	12.74	Unités et symboles
	01.78	Modificatif
X 02–002	10.75	Grandeurs et symboles
X 02–003	01.79	Grandeurs, unités et symboles. Principles d'écriture

British Standards (BS)

350 Conversion factors and tables

Part 1: 1974 Basis of tables. Conversion factors

Part 2: 1962 Detailed conversion tables (withdrawn 1981)

Supplement No. 1: 1967 (1982) to Part 2

3763:1976 The International System of units (SI)

5233:1975 Glossary of terms used in metrology

5555:1981 Specification for SI units and recommendations for the use of their multiples and of certain other units (=ISO 1000)

5775 Specification for quantities, units and symbols

Part 0: 1982	General principles (=ISO 31/0)
Part 1: 1979	Space and time (=ISO 31/I)
Part 2: 1979	Periodic and related phenomena (=ISO 31/II)
Part 3: 1979	Mechanics (=ISO 31/III)
Part 4: 1979	Heat (=ISO 31/IV)
Part 5: 1980	Electricity and magnetism (=ISO 31/V)
Part 6: 1982	Light and related electromagnetic radiations (=ISO 31/6)
Part 7: 1979	Acoustics (=ISO 31/VII)
Part 8: 1982	Physical chemistry and molecular physics (=ISO 31/8)
Part 9: 1982	Atomic and nuclear physics (=ISO 31/9)
Part 10: 1982	Nuclear reactions and ionizing radiations (=ISO 31/10)
Part 11: 1979	Mathematical signs and symbols for use in the physical sciences and technology (=ISO 31/XI)
Part 12: 1982	Dimensionless parameters (=ISO 31/XII)
Part 13: 1982	Solid state physics (=ISO 31/XIII)

Canadian Standards (CAN)

3–Z234.1–79	Canadian metric practice guide
3–Z234.276(80)	The International System of Units (SI)

French Standards

NF X 02–003 04/76	Principes de l'écriture des nombres, des grandeurs, des unités et des symboles
X 02–004 12/74	Noms et symboles des unités de mesure du système international d'unités (SI)
NF X 02–006 10/74	Le système international d'unités. Description et règles d'emploi. Choix de multiples et de sous-multiples

X 02–010	04/63	Sous-multiples décimaux du degré (unité d'angle)
X 02–011	11/74	Valeur de la pesanteur terrestre
X 02–012	03/77	Constantes physiques fondamentales
X 02–020	07/73	Définitions de termes liés à des grandeurs physiques
X 02–050	01/67	Principales unités de mesure américaines et britanniques
X 02–051	04/74	Unités de mesure. Facteurs de conversion
X 02–200	12/79	Grandeurs et symboles–Liste alphabetique
NF X 02–201	12/79	Grandeurs, unités et symboles d'espace et de temps
NF X 02–202	12/79	Grandeurs, unités et symboles de phénomènes périodiques et connexes
NF X 02–203	12/79	Grandeurs, unités et symboles de mécanique
NF X 02–204	12/79	Grandeurs, unités et symboles de thermique
NF X 02–205	12/79	Grandeurs, unités et symboles d'électricité et de magnétisme
NF X 02–206	12/79	Grandeurs, unités et symboles des rayonnements électromagnétiques et d'optique
NF X 02–207	12/79	Grandeurs, unités et symboles d'acoustique
NF X 02–208	12/79	Grandeurs, unités et symboles de chimie physique et de physique moléculaire
NF X 02–209	12/79	Grandeurs, unités et symbols de physique atomique et nucléaire
NF X 02–210	12/79	Grandeurs, unités et symboles de réactions nucléaires et de rayonnements ionisants
NF X 02–213	12/79	Grandeurs, unités et symboles de la physique de l'état solide

German Standards (DIN)

1301 T1	04.82	Einheiten; Einheitennamen, Einheitenzeichen

	Bbl 1	04.82	Einheiten; Einheitenähnliche Namen und Zeichen
1301	T2	02.78	Einheiten; Allgemein angewendete Teile und Vielfache
1301	T3	10.79	Einheiten; Umrechungen für nicht mehr zu verwendete Einheiten
1304		02.78	Allgemeine Formelzeichen
1304	Bbl 1	04.79	Allgemeine Formelzeichen; Zusammenhang mit internationalen Normen
1305		05.77	Masse, Kraft, Gewichtskraft, Gewicht, Last; Begriffe
1306		12.71	Dichte; Begriffe
1310		12.79	Zusammensetzung von Mischphasen (Gasgemische, Lösungen, Mischkristalle); Grundbegriffe
1311	T1–4	74	Schwingsungslehre
1313		04.78	Physikalische Größen und Gleichungen; Begriffe, Schreibweisen
1314		02.77	Druck; Grundbegriffe, Einheiten
1315		03.74	Winkel; Begriffe, Einheiten
1319	T1–3	71–80	Grundbegriffe der Meßtechnik
1320		10. 69	Akustik; Grundbegriffe
1324		01.72	Elektrisches Feld; Begriffe
1325		01.72	Magnetisches Feld; Begriffe
1326	T3	03.74	Gasentladungen; Benennungen, Formelzeichen
1332		10.69	Akustik; Formelzeichen
1338		07.77	Formelschreibweise und Formelsatz
1341		11.71	Wärmeübertragung; Grundbegriffe, Einheiten, Kenngrößen
1342		12.71	Viskosität newtonscher Flüssigkeiten
1343		11.75	Normzustand, Normvolumen
1344		12.73	Elektrische Nachrichtentechnik; Formelzeichen
1345		09.75	Thermodynamik; Formelzeichen, Einheiten
1349	T1	06.72	Durchgang optischer Strahlung durch Medien; Optisch klare Stoffe; Größen, Formelzeichen und Einheiten

1358	07.71	Meteorologie und Geophysik; Formelzeichen
4896	09.73	Einfache Elektrolytlösungen
4897	12.73	Elekrische Energieversorgung; Formelzeichen
5031 T1	03.82	Strahlungsphysik im optischen Bereich und Lichttechnik; Größen, Formelzeichen und Einheiten der Strahlungsphysik
5031 T4	03.82	Strahlungsphysik im optischen Bereich und Lichttechnik; Wirkungsgrade
5031 T8	05.77	Strahlungsphysik im optischen Bereich und Lichttechnik; Strahlungsphysikalische Begriffe und Konstanten
5476	04.78	Zeitbezogene Größen; Bilden von Benennungen
5483	02.74	Zeitabhängige Größen; Formelzeichen
5485	05.77	Wortzusammensetzungen mit den Wörtern Konstante, Koeffizient, Zahl, Faktor, Grad, Maß, Pegel
5490	04.74	Gebrauch der Wörter bezogen, spezifisch, relativ, normiert und reduziert
5491	09.70	Stoffübertragung; Diffusion und Stoffübergang; Grundbegriffe, Größen, Formelzeichen, Kenngrößen
5492	11.65	Formelzeichen der Strömungsmechanik
5493	08.72	Logarithmierte Größenverhältnisse (Pegel, Maße)
E 5493	10.80	Logarithmierte Größenverhältnisse (Maße, Pegel in Neper und Dezibel)
5496	07.71	Temperaturstrahlung
5497	12.68	Mechanik; Starre Körper; Formelzeichen
6814 T2	01.80	Begriffe und Benennungen in der radiologischen Technik; Strahlenphysik
6814 T3	06.72	Begriffe und Benennungen in der radiologischen Technik; Dosisgrößen und Dosiseinheiten
E 6814 T3	09.79	ditto
8941	01.82	Formelzeichen, Einheiten und Indizes für die Kältetechnik
13 320	06.79	Akustik, Spektren und Übertragungskurven; Begriffe, Darstellung

13 345	08.78	Thermodynamik und Kinetik chemischer Reaktionen; Formelzeichen, Einheiten
13 346	10.79	Temperatur, Temperaturdifferenz; Grundbegriffe, Einheiten
25 404	09.76	Kerntechnik; Formelzeichen
28 400 T1	07.79	Vakuumtechnik; Benennungen und Definitionen; Allgemeine Benennungen
28 402	12.76	Vakuumtechnik; Formelzeichen, Einheiten; Übersicht
32 625	07.80	Größen und Einheiten in der Chemie; Stoffmenge und davon abgeleitete Größen; Begriffe und Definitionen
40 110	10.75	Wechselstromgrößen
40 121	12.75	Elektromaschinenbau; Formelzeichen
45 630 T1	12.71	Grundlagen der Schallmessung; Physikalische und subjektive Größen von Schall

New Zealand Standards (NZS)

6501:1973	International System of Units (SI)
6502:1972	Metrication factors and tables for conversion to SI units (Revalidated 1979)

Swiss Standards (SNV)

012100–1978	SI–Einheiten und Empfehlungen für die Anwendung ihrer Vielfachen und Teile....
012110	Umrechnungsfaktoren für das internationale Einheiten-System (SI)

US Standards

NBS Special Publication 330:1981 The International System of Units (SI)

NBS Handbook 130:1982 Model State Laws and Regulations

International Standards

ISO 31/0–1981	(E)	General principles concerning quantities, units and symbols
	(F)	Principes généraux concernant les grandeurs, les unités et les symboles

ISO 31/1–1978 (E)	Quantities and units of space and time
(F)	Grandeurs et unités d'espace et de temps
ISO 31/2–1978 (E)	Quantities and units of periodic and related phenomena
(F)	Grandeurs et unités de phénomènes périodiques et connexes
ISO 31/3-1978 (E)	Quantities and units of mechanics
(F)	Grandeurs et unités de mécanique
ISO 31/4–1978 (E)	Quantities and units of heat
(F)	Grandeurs et unités de chaleur
ISO 31/5–1979 (E)	Quantities and units of electricity and magnetism
(F)	Grandeurs et unités d'électricité et de magnétisme
ISO 31/6–1980 (E)	Quantities and units of light and related electromagnetic radiations
(F)	Grandeurs et unités de lumière et de rayonnements électromagnétiques connexes
ISO 31/7–1978 (E)	Quantities and units of acoustics
(F)	Grandeurs et unités d'acoustique
ISO 31/8–1980 (E)	Quantities and units of physical chemistry and molecular physics
(F)	Grandeurs et unités de chimie physique et de physique moléculaire
ISO 31/9–1980 (E)	Quantities and units of atomic and nuclear physics
(F)	Grandeurs et unités de physique atomique et nucléaire
ISO 31/10–1980 (E) corr. 02–1982	Quantities and units of nuclear reactions and ionizing radiations
(F)	Grandeurs et unités de réactions nucléaires et rayonnements ionisants
ISO 31/11–1978 (E/F)	Mathematical signs and symbols for use in the physical sciences and technology
	Signes et symboles mathématiques à employer dans les sciences physiques et dans la technique

ISO 31/12–1981	(E)	Dimensionless parameters
	(F)	Paramètres sans dimension
ISO 31/13–1981	(E)	Quantities and units of solid state physics
	(F)	Grandeurs et unités de la physique de l'état solide
ISO 921–1972	(E/F)	Nuclear energy glossary
ISO 921–1972	(E/F)	Addenum 1–1980
ISO 1000–1981	(E)	SI units and recommendations for the use of their multiples and of certain other units
	(F)	Unités SI et recommandations pour l'emploi de leurs multiples et de certaines autres unités
ISO 1144–1973		Textiles – Universal system for designating linear density (Tex system)
ISO 3529/1–1981		Vacuum technology–Vocabulary

IEC Publications

27	Letter symbols to be used in electrical technology
27–1:1971	Part 1: General
27–2:1972	Part 2: Telecommunications and electronics
27-2A:1975	First supplement to 27–2
50(111–01):1982	Advance edition of the International Electrotechnical Dictionary, Chapter III: Physics and Chemistry, Section 111–01: Physical Concepts

British Acts and Regulations

Weights and Measures Act 1963 and 1976
The Units of Measurement Regulations 1976 and 1980

EEC Directives

Council Directive 80/181/EEC of 20 December 1979 on the approximation of the laws of the Member States relating to units of measurement . . .

French Index

This index sets out the French names of all the quantities and constants included in the body of this book. Names of units have not been included. The 3 capital letters are the first 3 letters of the corresponding English term in Part 2.

German Index

This index sets out the German names of all the quantities and constants included in the body of this Dictionary. Names of units have not been included. The 3 capital letters are the first 3 letters of the corresponding English term in Part 2.